UNCOMMON STRUCTURES, UNCONVENTIONAL BUILDERS

UNCOMMON STRUCTURES, UNCONVENTIONAL BUILDERS

Alan Van Dine

BLACK DOG
& LEVENTHAL
PUBLISHERS
NEW YORK

Copyright © 1977 by J. G. Ferguson Publishing Company
Copyright © 2001 Black Dog & Leventhal Publishers, Inc.

All rights reserved. No part of this book may be reproduced in any form or by any electronic or
mechanical means including information storage and retrieval systems without the written permission
of the copyright holder.

Published by
Black Dog & Leventhal Publishers, Inc.
151 W 19th Street
New York, NY 10011

Distributed by
Workman Publishing Company
708 Broadway
New York, NY 10003

Designed by Dutton & Sherman
Photo Research by Kimberley Mangun
Boxes by Jennifer Savage and Jenny Wierschem
Book Manufactured in Spain
DL: TO-699-2001

Library of Congress Cataloging-in-Publication Data

Van Dine, Alan.
   Uncommon structures, unconventional builders / by Alan Van Dine.
      p. cm.
   ISBN 1-57912-161-6
      1. Buildings—Popular works. 2. Antiquities—Popular works. 3. Animals—Habitations—popular
works. 4. Curiosities and wonders—Popular works. I. Title.

TH148 .V26 2001
690—dc21   2001018415

ISBN: 1-57912-161-6
h g f e d c b a

Dedicated to clever men

like Christopher Wren

Beowulf

Shakespeare

Joshua

Ch'in

Louis XIV

Incas

Eskimos

Irish monks

Neanderthals

ghosts

and clever birds

like the Least Bittern

Contents

Preface

The year Benjamin Franklin was born, a six-inch tooth was unearthed in Massachusetts. In those dim pre-Darwin, pre-paleontology days, the discovery seemed sufficiently ominous that the tooth was promptly brought before the Governor of the colony. Lacking certain essential points of evidence—for example, that mastodons ever existed—the Governor announced that the tooth belonged to a giant man who had been drowned in the Great Flood.

It is never easy to take a piece from the past and imagine its owner, its context, or its reason for being. When the remnants are the largest artifacts of man—his buildings—the imagining is easier but sometimes just as wild. And given the intensity of human hope, ambition, fear, greed, pride, or reverence necessary to prompt such an arduous project as the raising of a building, wild imaginings are often in order:

A terrified widow appeases the ghosts she fears by building them a 700-room house.

An emperor becomes so obsessed with national defense that he sets out to build a wall around his country.

To hide from predatory Vikings, Irish monks build a refuge tower visible for miles.

A city, bankrupt and starving after a ten-year siege, pours most of its marketable metals into construction of the largest, most extravagant statue on earth.

A small French town races to build the highest spire in history—so high and so hastily built that it cannot stand up.

An otherwise masterful politician commits political suicide by building himself a house that is better than the King's.

The pilgrim wades ashore at Jamestown and builds what? A log cabin? No, a wigwam, and then a medieval English cottage.

The record is not always clear as to why people build the things they do, in the way they do. Ireland's peculiar Round Towers stood in place so long that people in the surrounding countryside entirely forgot why or how the towers first appeared. After a thousand years, even direct descendants of the occupants and builders could be convinced that the towers had been erected in ancient times by visiting Carthaginians, Phoenicians, or Egyptians.

The Vikings, whose pillaging had prompted Irish Monks to build those towers in self-defense, left some architectural mysteries of their own. Why, for example, did these rough-cut, roving warriors trouble to shape the walls of their longhouses into sweeping, convex curves? There were clues in the sophisticated joinery of Viking structures. Though erratic as architects, the Norsemen were masterful shipbuilders. Some of their houses and halls had roofs that looked suspiciously like boats turned inside down, and the walls might simply have followed the boat-hull curvature of the roofing edge.

With a few exceptions the stories in this book have to do with structure rather far removed from the familiar monuments and the well-worn paths of tour guides and architectural texts. Libraries may bulge with explications of Shakespeare's plays, but his Globe Theater remains a mystery: no building, no ruins, no plans, no eyewitness descriptions have been seen for over three hundred years. The Tower of Babel has vanished, as has Heorot, the fabled mead hall where Beowulf struggled with Grendel. The log cabin is well known but widely misunderstood, thanks to the ballyhoo of two nineteenth-century presidential elections. Pizarro wrecked the Inca monuments. Louis XIV sacked the magnificent chateau Vaux le Viconte in order to embellish Versailles. And gravity felled the incredible spire of Beauvais Cathedral.

Any modern account of structures such as these—and the high hopes they once carried aloft—must begin with ruins and proceed through legend to what can reasonably be regarded as fact. In most cases scholars have sifted the evidence with more shrewdness and a great deal more data than the Governor could muster when presented with the remains of the mastodon. Each chapter in this book is indebted to researchers too numerous to credit who have done the real work of compiling fact and theory.

These narratives were commissioned by Koppers Company, Inc. and first appeared in a series called Tangents—a series of discussions for the building design professions, published by Koppers and recommended by them as non-urgent reading for architects and engineers. They are presented here for anyone who has ever stared at an odd-looking building and wondered why on earth anyone would want to build that.

UNCOMMON STRUCTURES, UNCONVENTIONAL BUILDERS

The remarkable 111-meter dome of St. Paul's Cathedral has survived fires, bombing raids and centuries of wind and weather.

Sir Christopher Wren

*Clever men
Like Christopher Wren
Only occur just now and then*[1]

I t has been said of Christopher Wren that never in his ninety-one years was he ever heard to express himself profanely. If that can be believed— that a man can practice architecture for more than half a century; can deal with five English monarchs, dozens of high officials of church and state, scores of committees and commissions, and hundreds of tradesmen and contractors; can settle thousands of questions of design, engineering, costs, and timetables; and can do it all without a single burst of profanity—well, if it can be said of

1. From verses by Hugh Chesterman, *Kings and Other Things*, Methuen, London

anyone, it can probably be said of Christopher Wren.

In speech and manner, he was by all accounts the most gentle of men. He had no taste for polemic, no penchant for assertion and rejoinder. The statements of Christopher Wren exist mainly in stone, in and about the precincts of London. His pen, though it seldom stopped for long, was reserved chiefly for his drawings and for the matter-of-fact correspondence necessary to see that the designs became buildings.

Words, after all, were not his primary medium. He had far more eloquent language at his command. Where exasperation might overtake others, Wren simply proceeded with the work at hand. His ideas were so compelling that they usually prevailed; when they were challenged, he would patiently explain his thinking. When even that did not carry the day, Wren would quietly *build* his arguments rather than continue to debate them.

The dome of St. Paul's stood tall and virtually unharmed after World War II air raids ravaged London in 1939.

Wren's rebuilding of St. Paul's Cathedral was completed 35 years after the confirmation of the building contracts, making it the first English Cathedral completed within the lifetime of the original architect.

Two stately columns—neither touching the ceiling they were supposed to uphold—stand as satirical essays on the foibles of committee design.

The Great Fire of September 2–5, 1666 destroyed three-quarters of the City of London, including Old St. Paul's Cathedral. Wren was appointed to the rebuilding commission and in the process of testing for rebuilding made discoveries concerning the archaeological remains of London.

Anticipating hard-core resistance to his idea for the great dome of St. Paul's, he calmly executed the same idea on a smaller scale in advance, at St. Stephen's, Walbrook. When decisions went against him, he would redesign, submit an acceptable compromise, and then later begin to make revisions that would restore much of his original plan. One controversy that went against Wren did evoke a touch of sarcasm—but even that was expressed in stone, and his irony proved so subtle that it was 250 years before anyone got the joke.

The occasion was Wren's election to Parliament in 1688, a compliment that he chose to repay by building a handsome town hall for the citizens of New Westminster. It was an excellent arrangement, a ground-floor arcade for the town offices and a large meeting room above. Everyone was happy except the mayor, who feared that the upstairs room was not

Sir Christopher Wren, shown here in a portrait from Wadham College at Oxford, was known not only for his architectural genius: He was also accomplished in geometry and astronomy.

adequately supported. Despite Wren's consummate mastery of mathematics, physics, and engineering, and despite his reassurances, the mayor continued to worry and finally convinced his official family that two more columns were needed to ensure their safety.

Wren the politician finally acceded to what Wren the architect knew to be redundant clutter. But he had the last word. Two and a half centuries later, workmen on high scaffolding saw what the mayor could not see from the floor: the two afterthought pillars rise almost—but not quite—

Uncommon Structures, Unconventional Builders

Decorated with frescoes and lined with windows, the dome and ceilings in St. Paul's lend an airy spaciousness to the interior of the building.

Wren's drawings are often just as beautiful as the buildings they portray. This drawing of St. Paul's Baptistry shows not only the exterior of the structure, but the interior and the foundations as well.

In his *Principia* of 1687, Sir Isaac Newton, who discovered gravity, ranked Sir Christopher Wren among the top three geometricians in the world.

to the ceiling. Two stately columns—neither touching the ceiling they were supposed to uphold—stand as satirical essays on the foibles of committee design.

Of course, in his day-to-day dealings not all of Sir Christopher's comments were so silent or so subtle. No architect completes seventy major buildings without some recourse to the ancient arts of presenting and explaining. Budgets must be coaxed into being; ruinous suggestions for "improvement" of a design must be neutralized; old habits of thought must be loosened soothingly or appeased with words before they are wrenched from their moorings by the massive audacity of some bold new structural form.

As the letters excerpted here show, Christopher Wren could soothe and

appease. He could explain, lecture, teach, and persuade. Far from being tongue-tied, he was a man of letters who could, among other things (among *many* other things), express himself lucidly in English and in Latin. He was of the age of the experimental philosophers, in the tradition of Sir Francis Bacon and the notion of the Renaissance Man—a scholar adept in every important area of human thought and art. And he was a genius.

He made fundamental contributions to astronomy, geometry, physics, even physiology. (He was the first to inject fluids directly into the bloodstream of an animal under laboratory conditions—opium, as it happened, in the vein of a dog who, sure enough, became sleepy.) He prepared a definitive treatise on spherical trigonometry.

Uncommon Structures, Unconventional Builders

He advanced astronomy with studies of the rings of Saturn and the moons of Jupiter, and with the design of instruments for observation. His inventions and theories—many of them since lost—included an elaborate weather clock, a device for writing in the dark, various musical instruments, a weaving machine, "divers new engines for raising of water," a method "for staying long under water," and another "to measure the Bass and Height of a Mountain only by journeying over it."

When he was invited to design his first buildings—including the Sheldonian Theatre at Oxford—it was not because he had formal training in architecture but because he had become known as one of the best minds in England. However, it was finally architecture that seemed to satisfy him. Once he began its practice, in his thirties, it became his life work, and he was soon the preeminent architect in England as well as the surveyor general, official architect for all projects under royal control.

Wren's importance and some of his exploits have nourished a popular conception of him as the aloof genius: the man who could majestically redesign all of fire-ravaged London before its ruins stopped smouldering (which he did) and who could later dismiss the demands of the world (which he didn't) with the oft-quoted:

"If anyone calls,
I'm designing St. Paul's."[2]

His actual days, as he lived them, were closer to earth and right in the thick of things. What the surviving records show is that designing buildings in the seventeenth century involved not only the same satisfactions but also the same day-in, day-out frustrations and reversals that every architect has wrestled with ever since—whether or not the architect is a supreme genius, and whether or not the client wears a crown.

Inklings of this are evident in some of Wren's correspondence, though only a small fraction of his notes and letters has been preserved. Fortunately there is one surviving series that tracks a design and construction project more or less from start to finish.

These letters are from Wren to Dr. John Fell, Anglican bishop and dean of Christ Church College at Oxford. Christ Church had the largest quadrangle in all

---

2. This quote, often attributed to Wren, actually comes from a piece of doggerel written about him by Edmund Clerihew Bentley:
Sir Christopher Wren
Said "I am going to dine with some men.
If anybody calls
Say I'm designing St. Paul's"

Above: The pineapple finial on the top of the front spire of St. Paul's symbolizes welcoming and goodwill. After Spanish explorers brought the exotic fruit to Europe from Central America, its likeness became commonly used in crests and architectural decoration to represent hospitality.

Below: Wren's 1669 model of St. Paul's is a work of art in itself—in the shape of a Greek cross, the model is topped with the impressive dome that was modeled after the Pantheon in Rome. The design was approved in 1675, but construction was ongoing until 1711.

According to Margaret Whinney's biography *Wren*, England didn't have a recognizable national style at the time that Wren began building.

The quadrangle at Christ Church College in Oxford was unfinished for many years, until Wren designed the magnificent tower gate.

The Sheldonian Theatre was one of the first buildings that Wren was commissioned to design. Patterned after the Theatre of Marcellus in Rome, the theatre follows classical Roman style with the exception of the innovative roof construction.

Sir Christopher Wren had no formal training as an architect.

of Oxford, thanks largely to the unceasing construction activity a century earlier under its founder, Cardinal Wolsey. Bishop Fell wished to finish what Wolsey had begun, and the notable piece of unfinished business was the main gate to the quadrangle. It had no gate tower, no proper house for Great Tom, the huge bell dedicated to St. Thomas of Canterbury and given to the college after its removal from Osney Abbey. Determined to build the tower, the Dean called on Christopher Wren at Wren's home in Scotland Yard.

Sir Christopher's first letter on the subject followed soon after.

> My Lord
> In pursuance of your Comandes I send your Lordship by Moores Coach my worke these Holydayes. I resolved it ought to be Gothick to agree with the Founders worke, yet I have not continued soe busy as he began. It is not a picture I send you nor an imperfect Essay but a designe well studied as to all the Bearing & fit to instruct the Workman if he will study the other Designe of the Groundplot, which at

first sight seems perplext, but he will find every necessary Line of several different plans (as the worke varies in rising) shewing how they beare upon one another, compare it with the orthography & the compasses will distinguish which is which plainer then words can expresse it.

The letter goes on to explain how Wren has resolved the structural problems of tying in the new tower to the older stonework and of supporting a square tower on an oblong gatehouse. He then suggests:

> Send me the rates & prices of Heddington, Burford Stone, Lime, with water & land carriages I can guesse what is fit to allowe for workemanship, & I will send Your Lordship a perfect estimate of the Charges, upon which you may contract to have it well and securely performed by X'Mas twelve-month.

He asks that his design be returned with the prices since there was only one copy but recommends that Bishop Fell have copies made for use in persuading potential contributors.

A point of concern for Wren at this stage is the likely quality of skilled labor the bishop will be able to obtain. "I cannot boast of Oxford Artists," Wren writes, "though they have a good opinion of themselves." His misgivings soon prove to be well-founded.

The first letter is dated May 26, 1681. Two weeks later trouble is already brewing, as Wren discovers the bishop's workmen have begun to lay foundation stones without waiting for the plan. The architect's letter of June 11 begins:

> The Designes (which I have not yet) shall not come sooner to my hands than Your Lordship shall have an estimate made and sent by the next post that followes, & and I am heartily sorry the foundations are begun without that due consideration which is requisite, soe that (unless you take them up again) I am out of all hope this designe will succeed. I am most certain it will not be without unexcusable flawes & cracks & weaknesse in the fabrick when the whole weight comes on, let your workmen warrant what they will.

After a patient explanation of how the foundation is to be revamped, using "the trew waies," Wren concludes:

> My Lord these & some other thinges in the consequence of the worke which are not provided by essential to the well performance, makes me jealous that your workmen beginning soe giddily will proceed accordingly & that you will find it too late, that every workeman is not fit for a great undertaking only because he is honest. My Lord if the foundations be only layd of Burres, I again entreat they be taken up, & the methods be used which I prescribe, & that your Lordship would pardon this freedome which in only zeale in.
> Your Lps most obedient humble sevant
> Chr. Wren

By late June there is a new cause for concern. Wren has now received from the dean a plan of the existing structure that differs radically from an earlier plan on which he had based his design.

> I thinke I have reckoned all particulars, according to my designes I sent, but I am confounded by the last peece of groundplot. I wrought by the first Groundplot which differe exceedingly from this. In the first I had 3 foot 9 inches between the inside of the inward Gate (measuring in the Torus's) & the inward angle; in this draught I have but 2 foot; in the first the College wall & the walls of the Gate are of one thicknesse 4 foot 3 inches. In this the sides of the porch are near 5 foot, & the College Wall not full 4 foot. If this be trew, I must correct many things in the Designe to adjust it as it should be, & I am uncertain whether I have the roome I desire for the peers that are to be added.

In the same letter, Wren instructs Bishop Fell on the methods of paying his stonemasons. He is careful to advise the dean to strike a good bargain but, at the same time, to make sure that the contractors are not held to unrealistically low estimates,

> ...for in things they are not every day used to, they doe often injure themselves, & when they are begin to find it, they shuffle and slight the worke to save themselves.

His recommendation:

> I thinke the best way in this businesse is to worke by measure; according to the prices in the Estimate or lower if you can, & measure the work at 3 or 4 measurements as it rises. But you must have an understanding trusty Measurer, there are few that are skilled in measuring stone worke, I have bred up 2 or 3. Mr. Kempster will wait on you the beginning of the weeke...

Uncertainties over the groundplot continue through the next letter (June 30), by which time Wren is requesting that his friend Kempster draw up a new one, "...soe from two witnesses wee shall find out the trewth." He asks the dean to halt all work until this is done. The confusion is causing

On the fire that destroyed the original St. Paul's, Wren wrote: "What time & weather had left intire in the old, & art in the new repaired parts of this great pile of Pauls, the late Calamity of fire hath soe weakened & defaced, that it now appears like some antique ruine of 2000 years standing, & to repaire it sufficiently will be like the mending of the Argo navis, scarce any thing will be left of the old."

He was of the age of the experimental philosophers, in the tradition of Sir Francis Bacon and the notion of the Renaissance Man—a scholar adept in every important area of human thought and art. And he was a genius.

delays, but Wren reminds his impatient client:

> I trouble your Lordship with scrupulosities but a thing well settled at first prevents more trouble, and errors in building are often incorrigible, & to peece well is a more carefull businesse than to erect a new thing where there is noe constraint, & that which is to last 500 yeares is not to be precipitated.
>
> You may doe enough this yeare to finish the next, & sooner will not be safe.

Next Wren's "perfect estimate" needs to be revised upward. The fourth letter is undated but apparently belongs to the summer months following the first three. By this time Kempster is on the scene and is supervising the stonemasons, which Wren predicts "will be apt to mend their pace." But the costs? Well,

> I have sent the articles, & because I find it hardly practible to drawe positive articles for every thing in such kind of worke as this is, but that there will be just reason to make small variations, & some things will be added to the worke as the hard stone, & some larger Heddington stone then Ashler, (which was before omitted) and more is taken down than I suspected at first to be necessary, upon these considerations I have added such Clauses as will keepe us from quarrelling at last, & between honest men I thinke if handes be set to the Articles as they stand it may suffice...

But the real problem is still to come. Since the bishop's letters did not survive along with Wren's, it's unclear how much of the cost revision was due to changes he insisted on, and it would be unlike the ever-tactful Wren to point the finger at his client unless the cost figures were challenged. The next change to be considered is clearly the bishop's idea, and it's a classic. Wren would need all of his tact and some very special expertise to deal with this one.

By December the work is progressing well, as Wren acknowledges in the first few lines of a letter dated Deceember 3. He does not mention how much of the structure is erected, but his original schedule to Bishop Fell in May had been, "...You may begin presently & rayse the peeres & tracery vault then let it rest and settle, let mortars dire, & finish next summer."

The bishop, however has had an inspiration. Sometime in early winter he ahs contacted Wren to propose that Tom Tower become not simply a bell tower over the gatehouse but, instead, *an astronomical observatory!* This, it occurs to him, might gain the college some leverage in getting contributions to finance the project.

If, in fact, his ninety-one years were free of profanity, this must have been one of Wren's closer calls. Suddenly his gate tower, begun "in the Gothick manner," had lost its point. Designed to rise to Gothic pinnacles and spire, it would now have to be contorted into a flat-roof building to accommodate an observatory.

The architect had spent his whole early life as an astronomer, and he had just finished the building of Greenwich observatory. That structure (which is

---

Opposite: Built in the 16th century, the lower portions of the buildings at Christ Church College in Oxford display ornamental Gothic style, but Wren combined elements of that ornamentation with plain surfaces and bold horizontal lines in the final design of the main gate.

Below: Wren constructed the octagonal tower at Christ Church College in stages. The cross-sections below illustrate how the form evolved from base to top.

Over the span of his career, Wren was the architect for a number of secular buildings in addition to his church work. Wren's first large secular building was Chelsea Hospital.

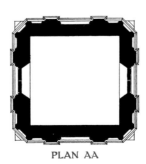

PLAN AA

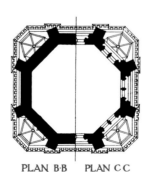

PLAN B·B   PLAN C C

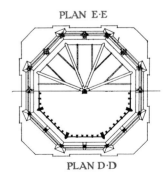

PLAN E·E

PLAN D·D

Architecture wasn't Wren's only strength. He was also an accomplished mathematician, astronomer and inventor.

The custom soon developed that when the big bell began to toll at nine each evening, students of the college knew they had exactly 101 bongs to get back into the quadrangle before the gates clanged shut for the night.

designed with a touch of the wizard's mystique, a sort of Merlin's castle) was envisioned, in Wren's words, "...for the Observators habitation & a little for pompe." The actual observations, he points out, are made not from the lofty platforms but from the court or the garden; the building is for apartments, offices, storage of instruments not in use, and, to suit His Majesty, "a little for pompe."

As for turning Tom Tower into an observatory, Wren seems at first to accept the notion and proceed to discuss what it will require to execute it.

> Yet I doe not reject your Lordship's proposition of making it into an Observatory, and I was willing first to get rid of severall businesses that incombred me at once the last weeke that I might at leasure consider of this change, for a change it will be of the whole designe.

As to Wren's "severall businesses, " it should be noted that at this time he had the building of St. Paul's Cathedral well under way as well as his famed Trinity College Library, Kilmainham hospital, and at least a dozen London churches, rebuilding after the Great Fire. And as Surveyor of the King's Works, he was constantly on call for consultations regarding every piece of royal construction or remodeling. Still his attention to Bishop Fell was thorough, deferential—at least to a point—and devastating.

Detailing the necessary changes, he reports:

> The loft for the Bell above the ringing loft must be higher considerable & with large Windowes & still I doubt the Bell will be somewhat lowe to be well heard; then the octagonall Tower must be flat on the top with a levell Ballaster (for pinnacles will doe injury) the windowes also must be only wooden shutters without Mullions for barres, these things considered it will necessarily fall short of the beauty of the other way, for

having begun in the Gothick manner wee must conclude above with flats & such proportions as will not be well reconcilable to the Gothick which spires upward & the pyramidall formes are essentiall to it, & this proposition had been much better effected had not the parts formerly built diverted us from beginning after the better formes of Architecture, & I feare wee shall make an unhandsome medly this way.

Wren goes on to describe how astronomical observations are made and with what equipment, pointing out that there is nothing about a tower which contributes to this work, "...for wee valewe noe observations made near the Horizon." He concludes the letter by promising to start over with the whole design if the bishop persists in his plan, but then adds a postscript which surely must have destroyed the whole idea once and for all:

> I cannot but adde thise Line that supposing all the necessary instruments provided for an observatory (& I dare undertake for the Charge of 300 lb or 400 lb at most to provide more usefull and accurate instruments than I find were ever yet made) *I will then for less charge than a pidgeon house provide all the housing necessary.*

Nine months later, Tom Tower stood complete, as originally designed, awaiting only a ball and vane for the top of the spire and the hanging of Great Tom itself. The custom soon developed that when the big bell began to toll at nine each evening, students of the college knew they had exactly 101 bongs to get back into the quadrangle before the gates clanged shut for the night. No one knows what those mighty vibrations might have done to the calibrations of astronomical instruments, but nearly three hundred years of wear and tear haven't opened a single "unexcusable" flaw or crack in Tom Tower, whose hard stones were, after some difficulty, "well peeced and set in the trew waies."

sept. 9th

My Lord

As soon as I receiued yr Lps first orders I put
in hand a Ball together with a vane & the Flowr
out of wch the stemme rises all of Copper, & the
spindle (upon wch the Ball is fixed & the vane
turnes) of Iron, wch to giue it good hold in the timber
worke I ordered to be 12 foot long, it will I hope
be finished in a weeke or 10 dayes, the Copper
worke may be sent by Waggon cased up, it
may be gilt there, if they will doe it as cheape
it will haue some care in the packing up, the
spindle (wch was don heer to fit it the better) if
it proue too long for land carriage, may be sent
by water while the Copper is in gilding. I thinke
the mason understands what to doe with the
Towers, I discoursed it with him, if he doubts
he shall haue farther directions. There was
a statue of the Kings to be sold if it be not
disposed of, I shall enquire. You will hardly
haue a statue for 200ll, for the brasse will be
80. of the mony. Cajus Gibber if I can find him
will be able to doe it as well, or I will enquire
farther, the Horse at windsor was first cut in wood
by a German, & then cast by one Beck, a founder
in London, but this is the dearer way, if wee
can find a good statuary for brasse it will be better.
Yr Lps most humble servant Chr. Wren

The Round Tower (900-1200) and church at Turlough, County Mayo.

Ireland's Curious Towers

*After a thousand years
no one could remember
how they got there*

Picture a medieval Irish monk... at a dead run. And sixty more or seventy: a monastery in turmoil, triggered by the tinkling of one small hand bell from the top of a tower. Utensils clatter to the floor in the scullery and the dining hall where the community has just finished breakfast. Someone is hurriedly stoking a tinderbox; others gather what they can carry of food staples and water. In the church sacristy, monks are hastily collecting prayer books, religious relics, and the sacred vessels. And out of the drafty hard bench halls, where reading, teaching, and the copying of documents had just begun for the day, students and their cleric teachers are running, their

arms laden with irreplaceable texts and manuscripts. Here at Ardmore in Country Waterford, as elsewhere in twelfth-century Ireland, virtually every scrap of written knowledge and tradition exists solely because the men of the monasteries have painstakingly recorded and preserved it.

But now they must save not only the manuscripts but themselves. The bell ringing is increasingly frantic. Some of the monks have reached the tower from which the warning signal sounded—a curious, pencil-shaped stone spire rising 130 feet to its conical belfry. From nearby farms people who have heard the commotion or sighted its cause are arriving, breathless and frightened, to join the monks. One by one they scramble up a ladder to the single entry—ten feet off the ground. Finally, everyone is inside, the ladder is pulled in after them, and the heavy wooden door slams shut.

Ireland's towers rise in rolling hills and pastoral countryside as a testament to Irish resourcefulness.

In a letter to the Viscount Adare and William Stokes, George Petrie paid tribute to Irish scholars. He said, "I would wish to be remembered hereafter, less for what I have attempted to do, than as one who, in the pure and warm hearts of the best and most intellectual of his local contemporaries, had found, and enjoyed, a resting-place, far superior to that of the Greek."

Uncommon Structures, Unconventional Builders

Opposite: Irish scholar George Petrie made the first systematic survey of the ancient towers in the mid-19th century. He was the first to note that the towers, like this one at Glendalough, are invariably found among monastic ruins.

Right: Windows served dual purposes in the towers; they let light into the dark rooms of the tower, and they provided a nice perch from which the monks could drop or pour things onto their attackers. Virtually every round tower has a window directly over the door, its sill sloping outward. Anyone trying to force through the door would have to stand in the bull's-eye position.

From the far side of the monastery, toward the bay, an awesome sight has emerged: the dreaded horned helmets, boar skins, heavy swords, shields, and bloodcurdling shouts in the unrecognizable tongue of the Norse invaders, landed in one of their countless raids on the villages and monasteries of England and Ireland. As a tribute to Odin, and to the ship's stores, they will sack the monastery. They will burn every book and parchment abandoned in panic. They will carry off whatever treasures have been unsuccessfully hidden. And, if they have the time and patience, they will outwait and possibly try to root out or burn out the refugees in the tower. But that will be a job.

The Round Towers of Ireland were so sturdy that eighty of them are still standing today, even where the churches and abbeys around them lie in rubble or have been leveled and overgrown for nearly a thousand years. The Towers did not succeed in saving Irish scholarship or in safeguarding any substantial percentage of the art and literature that flourished there despite the encircling Dark Ages. But they did save whatever *was* saved, and sometimes they saved the people who lived in constant peril from Viking raiders, who came and went at will for several hundred years.

Woefully cramped for refuge in siege, the Towers were nevertheless hard to breach. No warriors defended the portals—the occupants were conspicuously nonmilitary—but the single door was heavy, heavily bolted, and elevated beyond the reach of battering rams. The stone walls were at least three, sometimes five feet thick at the base. Windows were small, only one window per story, each facing in a different direction. Inside, there were from three to five floors, each a separate compartment connected to the one below only by a ladder, which could be retracted to the room above, and the

In the preface of his *The Ecclesiastical Architecture of Ireland*, Petrie explains the driving force behind his research. Petrie writes that he "was actuated solely to undertake this additional labour by an ardent desire to rescue the antiquities of [his] native country from unmerited oblivion, and give them their just place among those of the old Christian nations of Europe."

From the far side of the monastery, toward the bay, an awesome sight has emerged: the dreaded horned helmets, boar skins, heavy swords, shields, and bloodcurdling shouts in the unrecognizable tongue of the Norse invaders, landed in one of their countless raids on the villages and monasteries of England and Ireland.

Many of the towers withstood 300 years of Viking incursions and have outlasted their mother monasteries by nearly a thousand years more. Kilree Tower still stands, almost completely intact, among ruins of the ecclesiastical buildings it once served.

The 12th-century tower at Kilala in County Sligo.

Theories surrounding the uses of the towers were as scattered as the theories of their origin. Some theories claimed the towers were fire temples, places to hold Druid festivals, astronomical observatories, phallic emblems, Buddhist temples, anchorite towers, prisons, beacons and watch towers.

As for the towers, they often survived not only one raid but many, endured through 300 years of predatory Norsemen, finally outlasted the churches and monasteries they were built to serve, then stood for nearly a thousand years of disuse while the surrounding inhabitants entirely forgot how these curious structures ever rose to punctuate the Irish landscape.

When someone finally asked how the Round Towers came to be, the flowering of theories was remarkable. The Druids built them, one antiquary announced, as places from which to proclaim religious festivals. No, it was the Danes, or the Carthaginians, or the Phoenicians. The sloping window jambs of several towers led someone to connect them with the temples of ancient Egypt. Perhaps they were fire temples, possibly astronomical observatories, prisons, Buddhist temples, monastic castles, temples of Baal, beacons, watchtowers, stylite columns, or phallic emblems.

In 1845 Irish scholar George Petrie drove the bats out of the belfry with a careful, methodical analysis, establishing once and for all that the towers were Christian in origin (in fact, had Christian emblems in some cases), that they were always built in conjunction with monasteries, that their architectural features were unrelated to any remnant of pre-Christian structures in Ireland but closely related to their companion church buildings, that all were built during the period of Viking colonization and plunder, and that their uses included both the function of bell tower and that of refuge for "passive defence" in time of siege.

opening sealed to keep out attackers or smoke. Eventually, the invaders might burn out their prisoners, or starve them out, but the task was so lengthy and so difficult—while the main booty was in the monastery, not the tower, that the brothers very often survived the raids.

To that Margaret Stokes added the clincher in her 1878 volume, *Early Christian Architecture in Ireland*. She chronicled the Norse raiding patterns that inflicted up to twenty recorded raids on a single monastery between 800 and 1100. Then, having shown the need for a secure refuge to preserve men and property, she returned to the architectural evidence and made a shrewd series of observations about tower windows.

Although each window commands a different side, there is almost always one window directly above the door. And while the other windows may be placed high on the walls of their respective rooms for maximum light, the window over the door is close to the floor, and its sill slopes downward to the outside.

How that configuration would suit a Druid festival or the worship of Baal was hard to imagine—but it would certainly accommodate the dropping of heavy things on an attacker who was trying to breach the tower's only door.

Once this sort of systematic study had begun, other patterns emerged. All told, there were some 120 towers in Ireland, others in Scotland. Of the seventy-five or eighty survivors, twenty are almost totally intact. They range from 60 to 130 feet in height, from 40 to 60 feet in circumference, or only 13 to 19 feet in diameter. Each has from four to eight stories, roughly twelve feet in height. Demarcation between floors is often visible from the outside, either in the form of stone bands or of holes left in the walls to receive joists. Floors were usually wood, though a few were of stone, slightly arched.

The first towers were built of rubble masonry. Small stones, shaped by hammer, were fitted into the interstices between larger stones, all secured by lime mortar—a technique shared by many of the early Irish churches. Later examples, such as the Ardmore tower, show progress in masonry to fine ashlar work, along with a limited evolution in architectural nuance. Lintels, triangular arches, and primitive entablature for doors and windows gradually gave way to Romanesque arches as Irish missionaries returned from the European continent. Also, a modicum

of decorative work began to appear: corbels and carvings around windows and doors and in the belfries.

But compared with the intricate scriptural scenes and signs of the Irish High Crosses dating from the same period—in fact often sharing the same sites—the Round Towers remained plainly functional. Their roundness, their solid masonry bases, thick walls, slight taper, and double doors with heavy iron fittings all contributed to their amazing permanence through a period when virtually everything else was laid to waste.

Some were badly built. The tower of Kilmacdaugh in County Galway leans 17 ½ feet out of perpendicular, which is

Fine ashlar stonework distinguishes the Ardmore tower as one of the later examples of Tower design. Bands around the exterior indicate where the floors are inside the tower.

As many as 20 Viking raids hit a single monastery between the years 800 and 1100. Monks took refuge in the towers to save themselves and what every they could carry of the art and literature that would otherwise have perished during Europe's Dark Ages.

Their roundness, their solid masonry bases, thick walls, slight taper, and double doors with heavy iron fittings all contributed to their amazing permanence through a period when virtually everything else was laid to waste.

four feet further askew than the Tower of Pisa. Others simply fell down. But for a species born in a backward age of architecture—when what structural technology that did exist was oriented toward wood rather than stone—the Round Towers stand tall, each planted at the edge of an abbey graveyard but pointing in the direction singled out by the Irish monastic community as its purpose and destination.

The Round Tower in the village of Dervaig, on Ireland's Isle of Mull.

Opposite: When some of the original church and monastery stonework is still intact, it shows masonry methods very similar to that of the surviving towers.

Grave markers, grottoes and Irish crosses are usually among the ruins around Round Tower sites.

Uncommon Structures, Unconventional Builders

The Tower of Babel has been a favorite subject of artists for centuries. Here, the building of the tower is depicted in the Bedford Book of Hours, a 15th-century French work.

E-Temen-anki

The Tower of Babel

"Come, let us make a city and a tower, the top whereof may reach to heaven..."

Genesis xi:4

The Egyptians dug straight down and introduced the well. The Phoenicians headed out to sea and broadened the ancient world's idea of commerce. For the Sumerians the direction was up, and the result was the Tower of Babel.

The facts are buried forty centuries deep in history and lore. Just as no one is sure how deep the Egyptians dug, or how far the Phoenicians roamed, historians are uncertain as to how high the Sumerians were able to build. Still, the outlines of the story and the rough outlines of the tower have been pieced together.

Roughly 4,000 years ago the Sumerians migrated from an arid Iranian plateau to the vast, fertile plain of Shinar in Mesopotamia. Here, on the banks of the Euphrates and Tigris rivers, they began to farm, to quarry, and to build.

Aided by history's first standardized system of measurement—and spurred by the Sumerian belief that gods should be lodged far above the common mortal—they soon punctuated the plain with towering temples called ziggurats, jutting high above the flat city roofs to dominate a pilgrim's view from miles away.

By far the most conspicuous of the ziggurats stood in the city of Ka-dingir ("Gate of God"). When the city was conquered by the Semites, the name was literally translated to Bab-ilu. The tower was called E-temen-anki ("House of the Foundation of Heaven and Earth"). Translators of the Old Testament baptized the temple "The Tower of Babel."

Western scholars have been studying the ziggurat of Borsippa, just south of Babylon in Iraq, for a glimpse of how the Tower of Babel might have looked.

The Babylonians lived in Mesopotamia, a fertile land between the Tigris and Euphrates Rivers in present-day Iraq. Nimrod the Hunter was the first King of Babylon to commission the building of a tower to the sky. He was known not as a hunter of animals but as a hunter of the souls of men.

The theme of man as an overreacher—daring too much, challenging the gods, tampering with nature—has proved to be a persistent specter and one capable of many forms. Among artists, the Tower of Babel has served as a powerful symbol of this tendency, depicting the human attempt to reach into heaven without invitation. Here, 17th-century Dutch painter Tobias Van Verghaeght connects the sins of the past with those of the present by placing the tower in a busy seaport.

When translating the inscription, the Greeks used the word Babel for Borsippa, which means Tongue-tower. About 1,637 years after creation of the original tower it was rebuilt and an inscription written by the King of Babylon.

The Bible recounts how "brick for stone, and bitumen for mortar" were used to raise the tower heavenward, only to have the builders struck with "confusion of tongues" for their presumption.[1]

Whatever "confusion of tongues" may mean, there is at this point a confusion of legend. The most durable version of this story cast Babel in a role similar to Eve's apple: after challenging heaven with a tower, man was driven out of his linguistic Garden of Eden—where everyone spoke the same language—and was doomed to a world of many nations speaking many languages.

This, however, is not necessarily the biblical version; Genesis acknowledges the existence of different languages in the

---

1. A conflicting Mesopotamian legend tells how Nimrod, the Hunter (Gen. X:9) instigated a rebellion against the god who annihilated the world with a flood. Nimrod persuaded the rebels to build a tower high enough to escape the ravaging waters in case of another deluge.

chapter preceding the account of Babel. Nor is it the archaeological version. Scholars have established that at least two languages—Sumerian and Semitic—coexisted in Babylon prior to the tower.

It's possible that the entire "confusion of tongues" legend is traceable to a translator's error. The Akkadian word *bab-ilu* ("gate of God") is close to the Hebrew verb *balal* ("to confuse").

Whatever the case, the Tower of Babel *did* exist early in Babylonian history. Cuneiform texts, carved reliefs on ancient trays and urns, and detailed studies of the chronology of the Old Testament all attest that the great ziggurat was first raised as early as the Sumerians or their immediate conquerors.

But, as the war-torn Babylonian Dynasty faltered and eventually perished, the tower's sun-dried bricks fell prey to time and conquerors. (Some archaeologists contend that it was in ruins during Hammurabi's reign, around 1790 B.C.)

Uncommon Structures, Unconventional Builders

Mosaics at St. Mark's in Venice depict the tower's construction. According to Genesis, the people of Babylon said "come, let us build ourselves a city, and a tower with its top in the heavens, and let us make a name for ourselves." The Lord descended on their presumption by "confusing their tongues," making them unable to understand each other, thus unable to finish the tower.

In the Bible, Genesis 11:1-9 tells a story of how the people of the ancient world lived together and spoke one language. They built a great tower up to the sky that they hoped would reach the mists of heaven. God became angry with the people for trying to make a name for themselves. He confused their language and scattered them all over the earth.

The building of the Tower is often shown in Christian art, such as these stained glass windows.

And it was not until the reign of Nabopolassar (626-605 B.C.) that Babylon and its historic tower returned to glory.

Nabopolassar and Nebuchadnezzar, his son and successor, rebuilt Babylon on a spectacular scale. The city was crammed with wonders of the world: the celebrated Hanging Gardens...the garish Ishtar Gate with enameled-brick reliefs of bulls and dragons...brilliantly colored, 23-foot-thick walls that lined the procession path to the sacred zone of Marduk, the Lord of Eternal Life...and, most prominent, the detailed reconstruction of the original Tower of Babel. Once again, Babylon was the Capital City of the World.

Rebuilding the great ziggurat was the most ambitious task of all. His predecessors had made the mistake of using materials that quickly eroded; Nebuchadnezzar employed harder, properly fired bricks. Near the core of the eight-staged tower, the bricks were nearly forty-nine feet thick. According to one estimate, fifty-eight million bricks went into E-temen-anki.

The tower stood in a thick-walled, 500-yard by 450-yard enclosure, surrounded by small basilica-like chapels, pilgrim cottages, and a smaller temple called E-sagila, which housed a pure gold statue of the god Marduk. Herodotus, the Greek Historian, who visited Babylon, estimated the weight of the statue and other gold objects in the small temple at 800 talents. Estimated worth today: $85,000,000.

Information garnered from an excavated tablet shows that the Tower of

Babel had a square foundation, each side measuring slightly over 298 feet. Like an immense wedding cake, it rose in eight superimposed stages, each story smaller than the preceding one, to a height of 300 feet. Gigantic lateral stairways extended approximately one hundred feet up the tower to the first story, while another stairway, situated perpendicular to the face of the wall, led to the second story. From these points ascent was made by means of a series of secret ramps and

---

Artists have often depicted the tower as a tiered structure. Modern archaeology gives credence to this view—evidence suggests an eight-story building, each ascending level smaller than the one below.

Babel is made up of two Hebrew words. "Baa" means gate and "el" means God, thus translating to mean the gate of God.

Alexander the Great, leading his army through Babylonia on an expedition to India, ordered 10,000 of his soldiers to clear away the debris from the tower. He had to forsake the task eventually, but only after his army had spent a total of 600,000 workdays on the project.

The modern day Tower of Babel ruins, on the outskirts of ancient Babylon (modern Iraq), are currently 150 feet above the plain with a circumference of 2300 feet.

stairwells to the top. Couches for resting were positioned at various spots along the climb.

At the summit of E-temen-anki, Marduk resided in a small temple, or *shahuru*. The walls of this temple were gold-plated and decorated with bright blue enameled bricks. The room was empty but for a richly adorned couch (for Marduk to rest) and a gilded table. The only human permitted to enter this sanctuary was a woman, chosen by Marduk's priests to give the god "pleasure."

Imagine the sight that the multitudes of pilgrims beheld when they came to Babylon for the annual two-day procession for the god Marduk. As they entered the enclosure through Babu-Ellu ("Holy Gate"), they would worship the pure gold statue in E-sagila, go up the stairway and scan the entire city, craving their necks to

An engraving by the French printmaker Gustave Dore of the tower, a scene from Dore's Bible. After the original tower was destroyed, Nebuchadnezzar rebuilt the tower as part of an effort to raise Babylon up to be the most spectacular city on earth.

see the desert sun on the blue and gold walls of the *shahuru*.

However, the glory of the Tower of Babel did not last long. It did survive the city's capture by Cyrus the Persian (539 B.C.), who became so fascinated by the colossality of the structure that he ordered a smaller E-temen-anki be built for his grave; but the tower fell when Xerxes leveled Babylon nearly a century later.

Alexander the Great, leading his army through Babylonia on an expedition to India, ordered 10,000 of his soldiers to clear away the debris from the tower. He had to forsake the task eventually, but only after his army had spent a total of 600,000 workdays on the project.

Soon local inhabitants began carting away the hard building material for their own purposes, reminiscent of how the Vatican was built from the debris of Roman temples. Even today, a dam that diverts the Euphrates river into a canal is made of bricks that bear Nebuchadnezzar's royal stamp, while the site of the most famous religious tower in antiquity is nothing but a watery hole in the ground.

Right: According to legend, Marduk, the great God of Ur, resided in a small temple at the pinnacle of the great tower.

Below: In reconstructing the tower, Nebuchadnezzar ordered that only hard, kiln-fired bricks be used, so the tower would last forever. It lasted until Xerxes leveled Babylon but the bricks bearing Nebuchadnezzar's stamp can still be found, reused in house walls and a dam on the Euphrates River.

Dutch painter Hendrick van Cleve's (1525-1589) rendering of the fiery destruction of the tower.

The Winchester Mystery House is now one of the world's most famous haunted houses and a curiosity to historians and ghost-seekers alike.

Designs for the Haunted House

When the architects are ghosts the house is, of course, a nightmare

It ought to be possible to design a house in such a way that its occupants never see a ghost. The techniques for baffling, frightening, and denying access to would-be haunts are neither precise nor entirely consistent, but they are abundant.

What is vastly more difficult, judging from the ghostly lore, is the deliberate effort to design a haunted house—to achieve the character, the ambience, and the specific features that will invite ghosts to appear. It has been seriously attempted at least once, and that attempt took thirty-six years of construction and an investment of at least $5 million.

Money was no problem for Sarah Pardee Winchester. As a young woman, she inherited all the wealth generated by the invention and manufacture of the Winchester repeating rifle. Her problem, as she saw it and as an eager Boston spiritualist quickly verified, was the likelihood of vengeful ghosts.

The rifle had dispatched thousands of unwilling souls into the spirit world, all prematurely, none happily. If they came back, they would come back in unpleasant moods, and they would include in their number some very unpleasant types: frontier rowdies, murderers, bandits, drunks, and—most terrifying of all in Mrs. Winchester's mind—Indians!

With an ample construction budget and a retinue of willing spiritualists to advise her, Mrs. Winchester could have commissioned the design of a virtually foolproof, ghostproof house.

Mrs. Winchester added more than 700 rooms to the original 9-room house over a period of almost 40 years.

Sarah Pardee Winchester's house is a popular tourist attraction in San Jose, California. The Winchester Mystery House, as it is now called, offers daytime tours year-round as well as nighttime flashlight tours leading up to Halloween and on every Friday the 13th.

Sarah Pardee Winchester spent most of her life in flight from the evil spirits that she believed would follow her forever.

Winchester's fear was not only that vindictive ghosts would haunt her while she lived but also that they would ostracize her in the hereafter.

For example, although it is not clear why ghosts hate mirrors, it seems very clear to ghost watchers that they do. Confronted by a mirror, the apparitions vanish—some say because they are terrified by their own reflections; others say because ghosts are embarrassed or offended by the fact they have no reflections.

A house built with mirror walls in all rooms should be entirely free of hauntings. Short of that, a house designed without chimneys would deprive many ghosts of their favorite access and avenue of escape. A house surrounded by a moat is safe from the large majority of ghosts who can't or won't cross water. Towers and balconies should be eliminated; and since ghosts worry about time but depend on earthly timepieces (they normally arrive at or after midnight and must depart by two), clocks are taboo and bell towers are out of the question.

But to Mrs. Winchester the problem was not so simple. Her fear was not only that vindictive ghosts would haunt her while she lived but also that they would ostracize her in the hereafter. At the same time she had been assured that there were good ghosts as well as bad ones, and that the better spirits, properly appeased, would protect her in life and welcome her into polite society after death.

Through one of the spiritualists she also received word that no harm would come to her as long as builders were busy providing accommodations for the friendly spirits who would keep unfriendly spirits away.

So Mrs. Winchester began to build her house. From Boston she moved to California in 1884, and there bought a nine-room ranch house near San Jose. She hired twenty carpenters, soon added masons, glaziers, plumbers, and eventually electricians, and seven Japanese gardeners whose task was to raise and maintain a high hedge that would wrap the

More than 20,000 gallons of paint would be required to paint the house.

place in total privacy. Shifts were arranged so that the sound of hammers would continue year-round and around the clock.

Design conferences took place in the Séance Room, where Mrs. Winchester sat each evening, apparently alone. Her ghostly consultants were numerous but untrained, capricious, often vague, and utterly insatiable—demanding room after room, balcony after balcony, chimney after chimney. No mirrors. No moat. Forty-seven fireplaces, dozens of clocks (including three of the costliest chronometers in existence), watchtowers, a bell tower, and such personalized features as a floor made of seven kinds of hardwood and a stairway with seven turns and forty-four risers to ascend only nine feet.

The place was being designed by ghosts and for ghosts. Each morning plans drawn the previous night were handed to the head carpenter by Mrs. Winchester's personal secretary. No one else was permitted to see the widow. (Teddy Roosevelt once paid a call and was refused an audience. When she moved around the house, Mrs. Winchester wore a veil, and when two workmen accidentally saw her without it, they were discharged with a year's pay.)

At first these makeshift blueprints were nearly indecipherable and often unbuildable. As time went on, however, the drawings improved—even if their logic did not. To nine rooms were added dozens more, and finally hundreds more, many of them soon ripped out to make way for new ideas from Mrs. Winchester's nocturnal advisors. From its original site the house soon spread, crawling across a meadow to surround and swallow a nearby barn. A watchtower would rise, only to be engulfed by new layers of additions until its entire view was blocked.

Today, 160 rooms of this baffling labyrinth still stand, the survivors of an estimated 750 chambers, interconnected—if that is the word—by dead-end stairways, trick doors, self-intersecting balconies, and one stairway that goes down to a landing from which there is no other course except another stairway going up.

Uncommon Structures, Unconventional Builders

What happened in that Séance Room night after night is a secret buried with the widow, but if she was convinced that her ghosts were passionately particular about design details for the building, then she was dealing with a most unusual species of spook.

When Mrs. Winchester died in 1922, it took six weeks just to remove the furniture, partly because movers were continually going through one-way doors, finding themselves trapped and then lost for hours on end as they followed stairs to blank walls and corridors to unopenable doors.

Of her $21-million inheritance the widow had spent at least $5 million to please her discorporate friends. Unless ghosts are unspeakable ingrates, Mrs. Winchester has been well received and well regarded ever since.

If the building of the Winchester House seems eccentric to persons who do not take ghosts seriously, it must seem even more bizarre to those who do—or to the ghosts themselves.

In the thousands of recorded ghost appearances and the probably millions of

Although much of the house's architecture is designed to confuse ghosts, it is also a spectacularly lavish example of Victorian architecture. Parquet floors, elaborate stained glass windows, and exquisite craftsmanship can be found throughout the house.

Winchester rifles are now manufactured by the U.S. Repeating Arms Company under license from the Olin Corporation.

There are no master blueprints of the
Winchester House. It was built and rebuilt
from séance to séance, new plans materi-
alizing each morning. Mrs. Winchester
sketched instructions for the carpenters,
masons and glaziers on scraps of paper or
the séance room tablecloth.

The house was at one time seven stories tall but was knocked back down to four during the 1906 earthquake, which also left Mrs. Winchester trapped in a bedroom for hours.

After Mrs. Winchester's death, it took six weeks for movers to remove the furniture. They frequently found themselves trapped behind one-way doors, lost in labyrinthine corridors and frustrated by endless, aimless staircases.

unpublicized reports, the one reliable common denominator is that ghosts of the departed are seen and heard where the departed actually departed.

A Plains Indian might return to ride where he once rode, to hunt in the hills where he once hunted, to reenact the murder or the battle that did him in. But travel to California? Track down the whole chain of cause and effect from bullet to rifle to factory to owner to heir, and then leave his favorite haunts for a strange town? It would be most unghostlike.

By the established norms of haunting, all Mrs. Winchester had to do was determine that the original nine-room house she was buying (a) had not been haunted before, and (b) did not belong to General Custer.

What happened in that Séance Room night after night is a secret buried with the widow, but if she was convinced that her ghosts were passionately particular about design details for the building, then she was dealing with a most unusual species of spook.

It's true that there are creaking old manor houses which appear perfect for haunting. There are castles in England, Scandinavia, France, Germany, Italy, and Spain where ghosts are nothing short of mandatory, especially when the time comes to attract tourists. People respond—albeit some people more than others—to a sense of strangeness, of mystery, most often in dark, damp, drafty, and preferably very old dwellings. And certain architectural features seem to heighten this susceptibility to haunting: windowless attics, long corridors, cavernous basements, steep gables, belfries, balconies, towers, widow's walks, dramatic staircases, lone and lofty windows, and deep, bat-sheltering eaves.

But the ghosts themselves show a profound indifference to any such distinctions.

Their choices of places to appear are almost without discrimination as to architecture. They are reported not only in deep-dungeoned castles and ornate Victorian mansions but just as often in simple farmhouses, barns, box-Colonial subdivisions, and apartments—even in a cinderblock army post building in Texas and a topless nightclub in Toronto. The one possible exception is that it is very hard to find stories of haunted high-rise office buildings.

What matters to the ghost, in the great preponderance of reported cases, is what happened to him, not where or in

Patterns of 13 occur often throughout the design of the house. For example, there are 13 bathrooms, many rooms throughout the house have 13 windows, and 13 California Fan Palms line the front driveway.

what surroundings. A child who has witnessed a grisly murder returns to the scene. A man hanged by mistake lingers to clear his name. In most accounts of hauntings that claim to be true, the spirit comes back to erase, reenact, avenge, or simply continue to brood about some awful event or unfulfilled longing.

If so, the only ghosts likely to care a great deal about architecture would be the ghosts of architects. The notions that people have of what a haunted house is supposed to look like seem to spring from the fears of the living, their responses to unfamiliar surroundings, and from design decisions made not by architects or by ghosts but by writers: Edgar Allen Poe and kindred spirits fueling two hundred years of public enchantment with mysterious tales and the so-called Gothic novel. These and latter-day tendencies, such as the merchandising of Halloween, seem to have frozen a haunted house stereotype that the haunts themselves do not recognize.

The stairs in the Winchester House may be the most baffling feature—both for ghosts and people. Some of them ascend to blank ceilings; others lead to more stairways that go back down. This example takes seven turns and 44 steps to ascend only 9 feet. Mrs. Winchester avoided the confusion by simply taking one of the house's five elevators.

What matters to the ghost, in the great preponderance of reported cases, is what happened to him, not where or in what surroundings.

Not far from the haunts of Rip Van Winkle and Ichabod Crane, the U.S. Military Academy at West Point has had its share of Hudson Ghosts. The shape of a 19th century maid named Molly is said to rumple the bedclothes in the superintendent's mansion. A woman who died in a nearby house frightens guests by floating in through an upstairs window. Two cadets in the 47th division barracks reported several visits by a 19th century cavalry officer in full regalia.

Above: Nathaniel Hawthorne's House of Seven Gables stands under the curse of a wizard who was executed during the Salem witch trials.

Left: Netley Abbey, founded in the 13th century by Cistercian monks. In 1700, Walter Taylor bought the Abbey for its stone, but refused to set foot inside until the roof had been removed, claiming that the ghost of a monk had warned him that the roof would fall and kill him. After the roof was removed, he finally went to the Abbey, where a stone from high above the west window fell and fractured his skull.

With its 800-year history of murders and executions, the Tower of London has more cause for haunting than most places. Among other reported apparitions, visits by two executed wives of Henry VIII have gained the most notoriety. Katherine Howard has appeared, and in 1864 an iridescent Anne Boleyn approached a guard. The guard first tried to bayonet her, then fainted and was court-martialed for sleeping on duty. He was acquitted when the commander of the guard reluctantly admitted that he had been at an upstairs window and had seen the entire incident, ghost and all.

The Solar Observatory at Machu Picchu, because of its lofty elevation, is often surrounded by clouds.

Machu Picchu

Lost city of the Incas

The role of scapegoat in history, before it was preempted by politicians, was occupied largely by architects and writers. When war, disaster, or disease left a culture wondering whom to blame, the usual response was to raze the architect's monuments and temples and burn the writer's chronicles. And when invaders judged a conquered people to be wrong-headed or lazy, the army straightaway destroyed the books and buildings around which the offending culture was patterned.

Legend has it that the Incas once possessed a written language but, regarding it as the cause of a great pestilence, banned its use under pain of death and forbade even the traffic in *quilcas*—the parchment or leaves used for writing. This one was on the writers; the next one was on the architects: Pizarro's conquistadors, confronted by a recalcitrant non-Christian population, destroyed its temples and leveled its cities.

For hundreds of years, the known remnants amounted to scattered fragments of a once-magnificent architecture and a few, highly suspect written accounts.

Historians nourished themselves mainly on W.H. Prescott's famous chronicles, which were based on the royal commentaries of Garcilaso de la Vega, the son of an Inca princess and a Spanish conquistador. But Garcilaso was making the most of dim memories. He had left Peru as a teenager and did not begin to reminisce until he was an old man in Spain.

Another account, compiled by Fernando Montesinos, was somewhat clouded by the author's belief that Peru was originally developed under Ophir, the great-grandson of Noah. And, lacking a written language, the Incas themselves were unable to set the record straight with any obliging equivalent of the Rosetta Stone or the Dead Sea Scrolls.

Hiram Bingham, the Yale University professor who found the ruins at Machu Picchu in 1911, took a number of photographs during his survey and labeled each structure at the site. Bingham's note on the back of this photo, which was taken on July 25, 1912, indicates a "high-gabled double house #4001 and the S. corner of #4003. H.B."

Above: The mountain-locked citadel of Machu Picchu is built on staggering slopes. The Incas used sophisticated terracing to accommodate the pitch on which they were building.

Right: Hiram Bingham on his Peruvian Expedition of 1912.

Until 1948 the only way to reach Machu Picchu was by a 27-mile footpath. In 1948 the Peruvian government built the first road to the sacred ruins.

Into this vast historiographic vacuum trekked Hiram Bingham, in 1911, with a Yale expedition mounted to track down the persistent legend of a "Lost City of the Incas." Bingham's party sifted through rumors, pored over ancient maps, and began a ghastly trek down the Urubamba river, a headwater of the Amazon. Camped beneath a towering mountain called Machu Picchu, they heard that there were "interesting ruins" above. By this time the rigors of

jungle travel had dampened some of the group's exploring zeal. The expedition naturalist could see no reason to climb a mountain when he could, at leisure, find more interesting specimens near the river. The surgeon wanted to wash his clothes. Bingham finally went up with a Peruvian army sergeant and found what is generally regarded as the most important archaeological discovery in the Western Hemisphere.

Here were a hundred, ten-foot-high stone-faced terraces—each hundreds of feet long. Here was a cave lined with beautifully cut stone—apparently a mausoleum for royalty. Here was a great curving wall made from fine-grain ashlars of pure-white granite—more magnificent than any of the Cuzco walls which had been admired for centuries. Here was a flight of steps hewn from a single giant block of granite.

Here were houses with architectural details unique in all the world: stone pegs carved into the gable ends to secure roof thatching, eye-bonder courses drilled with sand and water to secure the roof purlins, massive bar locks carved from ten-ton doorway posts to secure the buildings against intruders. Here, without question, was the Lost City of the Incas.

In the view of its sixteenth-century inhabitants—and of today's scholars—Machu Picchu was fortunate to be "lost." Pizarro's conquistadors found Cuzco, the largest and most resplendent city of pre-Columbian America, and demolished it. They found Inca cities, Inca temples, and a luminous Inca culture throughout vast stretches of present-day Peru, Ecuador, Bolivia, Argentina, and Chile—and reduced it all to an imporverished serf-

dom. But in forty years of searching, they could not find Machu Picchu; so this remarkable monument to Inca genius survived the coming of "civilization."

Once Bingham had succeeded where Pizarro had failed, the evidence of Machu Picchu combined with other Inca ruins to reveal a startling record of architectural and engineering accomplishments. Examples:

The Incas were great irrigators. Without surveying instruments, they ran perfect contour lines dozens of miles through the valleys, and down from some

By domesticating llamas and alpacas—both mountain relatives of the camel—the Incas probably saved these animals from extinction. Smaller than the llama, the alpaca was bred by the Incas for its fine wool. Modern Peruvians, as did their Inca forebears, rely on the assistance of llamas and alpacas to negotiate loads through the rocky Andes.

Near Cuzco there is a fort with walls that contain 300-ton blocks so perfectly mated that the cracks between them will not admit a knife blade.

Machu Picchu is Peru's largest tourist attraction with trains running to the site. The Peruvian government has endorsed plans to run a tram to the top of the ruins. It will allow 400 tourists an hour to visit the site. Local citizens and international heritage groups have launched an education campaign in hopes of urging government officials to reconsider.

Farming terraces and stone buildings were connected by stairways carved into the granite cliffs. Aqueducts provided this self-sufficient city with water for drinking and irrigation.

of the highest, most rugged mountains in the world. Aqueducts supplied the cities.

The Incas were prodigious wall builders. Near Cuzco there is a fort whose walls contain 300-ton blocks so perfectly mated that the cracks between them will not admit a knife blade. Without mortar, without clamps, without iron or steel tools, the Incas shaped these huge pieces by hammering them with rocks and honed smaller blocks with abrasive sand mixtures.

The Incas were great temple builders. When Pizarro's men invaded Cuzco, they tore down the mammoth Temple of the Sun and erected a church on its foundation. The church has to be rebuilt after each earthquake, but the foundation blocks are as tight as ever.

The Incas were great road builders. They laid paved roads from Ecuador to Chile and from the mountain slopes to the Pacific. Typically five feet wide, the Inca roads are made of smooth rock slabs—sturdy enough pavement that much of it is in perfect condition today. It curls around 65-degree mountain precipices, supported by well-drained stone walls. It has steps where the grades are steep and, where necessary, tunnels through solid rock. When Pizarro arrived, thousands of miles of paved roads lay waiting for the invention of the wheel.

The Incas were daring bridge builders. They liked to live high, like eagles (Cuzco,

Uncommon Structures, Unconventional Builders

Spanish conquistadors thought the streets of Machu Picchu were paved with gold.

Left: The Incas were remarkably adept at mortarless masonry, relying instead on joinery and stones that fit together tightly.

Below: The Sun Temple of Torreon is the only circular structure at Machu Picchu.

Tucked away high in the Andes, its location a secret and its existence a mere rumor, Machu Picchu escaped Pizarro's devastating conquest of Peru and remained an unconfirmed legend until early in the 20th century.

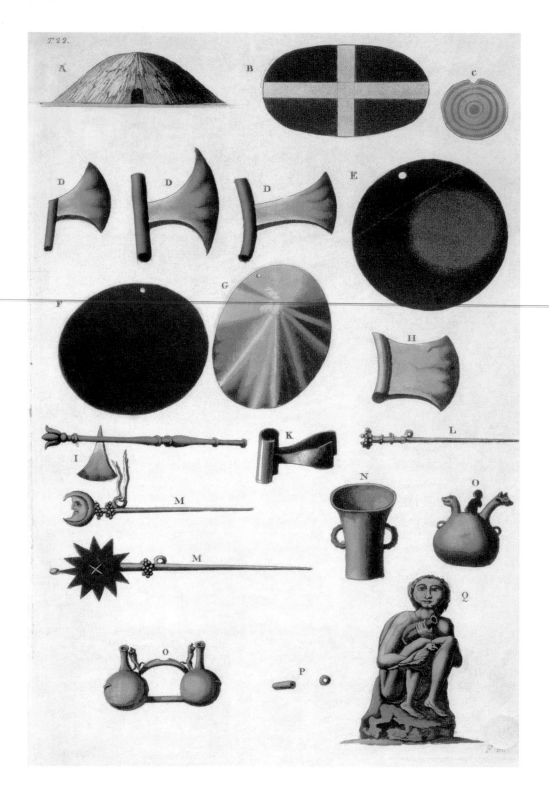

The Incas long recognized the value of fertilizer, and certain offshore guano-producing bird islands were allocated to various provinces. It was forbidden to visit these islands during mating season; and it meant the death penalty to kill even one of the hundreds of thousands of guano-producing birds.

woven from liana vines. To destroy a bridge meant the death penalty; and this, in a way, contributed to the downfall of the Incas. When the Spaniards were in hot pursuit, it never *occurred* to the Incas to destroy their bridges behind them.

The Bronze Age lingered late in Peru and environs, but Inca achievements are impressive nonetheless—not only in architecture and engineering but in medicine, agriculture, and animal husbandry as well:

It is estimated that the Incas domesticated more food and medicinal plants than any other people of the time. For example, they found a pea-sized edible tuber high in the Andes and cultivated it into what is now know as the Irish potato. After 300 years and two major famines, Europeans discovered the value of the potato, and today there are Sherpas in highest Nepal living on the same plant (they got it from the British Embassy garden).

The Incas also developed the sweet potato, which they called the *cumara*. Today it is cultivated all over the Polynesian islands, from Hawaii to New Zealand, and it still has roughly the name the Incas gave it: *Kumara* or *Kumala*.

The Incas long recognized the value of fertilizer, and certain offshore guano-

With nothing more than stone implements like these, the Incas were able to build interlocking walls of perfectly carved and fitted stones. The stones, quarried from the mountaintops, had to be shaped to fit sophisticated patterns and curvatures. The blocks, some weighing many tons, were probably beaten and chipped to approximate shape with flint or stone tools, scraping and abrasive grinding with sand mixtures.

for example, is above 10,000 feet), apparently because every venture into the valleys ended in warfare with less civilized local tribes. So they became bridge builders of most magnificent mien, spanning river gorges with 300-foot suspension bridges

The buildings of Machu Picchu are all made of local stone. Various walling materials, including coursed ashlar to roughly dressed rubble can be seen at the site.

Uncommon Structures, Unconventional Builders

The Incas crossed mountain slopes with complex irrigation systems, roads, bridges and aqueducts to cultivate crops.

Left: Trapezoidal windows are common in the ruins, which are extraordinarily intact, given the centuries of earthquakes in the area.

Below: Had the Spanish Conquistadors found this fabled city, they would have been disappointed; the streets were not, after all, paved with gold. The central plaza and Huayna Picchu are, however, remarkable in their own right.

Machu Picchu: Lost city of the Incas

producing bird islands were allocated to various provinces. It was forbidden to visit these islands during mating season; and it meant the death penalty to kill even one of the hundreds of thousands of guano-producing birds.

Masters at domesticating animals, the Incas tamed a small, hard-to-catch Andean rodent called the *cuy*. Today it's known as the guinea pig, although it does not come from Guinea and is not a pig. Even today when an unexpected guest arrives at an Indian home in Peru, the friendly little creatures are scooped off the floor and made into cuy stew—acclaimed as a delicacy by the most sophisticated gourmets.

The Incas also domesticated two species of the wild American camel—the coarse-hared llama as a beast of burden, and the silky-haired alpaca for its fine wool.

Under the conquistadors, the demise

Most of the rooms of Machu Picchu are one-room stone houses arranged around an internal courtyard.

of the Incas was tragically swift. The original conquest was made by barely 200 men, but they were aided by an Inca civil war, by the terrifying sight of the huge Spanish horses, and by the crack-of-doom sound of Spanish firearms.

The final blow was struck by the Spanish viceroy Francisco de Toledo. To avenge the murder of an ambassador and a priest, and in a last frantic search for gold, he ordered his soldiers in total pursuit of the Inca emperor Tupac Amaru. They caught him in the Amazon jungle, returned him to Cuzco, and executed him—along with his entire family. Thus, in 1571, less than 40 years after the conquistadors arrived, the last Inca emperor died, and the great civilization sank to its knees for all time.

The conquistadors took many things from the Incas: their land, their freedom, their religion, their gold, their lives. But the ghosts of the Incas may be laughing last. They never yielded Machu Picchu. And they gave the conquistadors one more prize of conquest than they bargained for. The Spaniards returned home not only with Peruvian gold but also with Europe's first major infusion of syphilis.

Today, indigenous Peruvians in traditional dress can be found at Machu Picchu for the benefit of tourists. Thousands of interested travelers visit the ruin each year—a fact that has prompted discussion about the impact of tourism on the ancient city.

Opposite: The view from the ruins is no less than spectacular. The Incas were not daunted by the harsh geology at such a high altitude.

In 1571, less than 40 years after the conquistadors arrived, the last Inca emperor died, and the great civilization sank to its knees for all time.

The sparrowlike weaverbird family
includes some of the most accomplished
roof-thatchers in the business. The masked
weaver, shown here, weaves a nest that
ends up looking like a ball of yarn.

The Cuckoo Construction Company

How to design a home when the most important thing you'll ever own is an egg

With a spare, muscular body, a strong heart, perfect digestion, rich blood, and incredibly vigorous metabolism, a bird has no excuse for being less than an energetic builder. And given his need for a safe, soft, warm, protective residence in which to conduct a precarious hatching and raising of children, the bird has ample incentive to build the perfect nest. Add 150 million years of evolution to develop the instincts of construction (the beaver has had less than one million), the same eons for natural selection to weed out defective nest-building techniques, and the practice-

makes-perfect benefits of having to structure a new home every year—add it all up, and you might expect somewhere to find the equivalent of Gold-Finch Gothic, Bluebird Baroque, Redstart Renaissance, or Warbler Williamsburg.

It didn't work out that way. Nature is awesome in many respects but not all. The ostrich is one of several birds that can't fly. The chicken stays out in the rain and drowns. The squirrel forgets where he buried his nuts. And when it comes to nest building, it would have to be said that in many cases Aves doesn't try hard enough.

The grouse scratches out a little dirt, pulls in a few leaves, and lays its eggs in this shallow depression completely accessible to predators. The yellow-billed cuckoo makes a skimpy platform of twigs and straw so haphazardly thatched that the eggs occasionally plummet right through

Four cliff swallows walk through the mud looking for nesting materials in the Bear River National Wildlife Refuge in Utah.

Some swifts are interesting for their edible architecture: bird's nest soup is made from the salivary secretions used by some species of the Oriental swiftlet in building their nests.

71

Swallows are common—and noisy— inhabitants of North American and Western European barns.

it. Some birds hardly even make an attempt. The band-tailed pigeon, whose single white egg would seem to deserve high priority protection, either throws together a loose, jackstraw assembly of twigs or ignores even that formality and lays the egg on the forest floor. The English kestrel will settle for an abandoned crow's nest. And the cowbird simply lays her eggs in other birds' nests and leaves the hatching to them.

In fairness to birds and in quaking respect for bird-lovers, it should be conceded that avians do tend to build roughly what they need. If the resident is weatherproof, the house needn't be. Nests are for nesting, for incubation and feeding—for eggs and their occupants— and in accommodating this cargo, birds' nests reveal a construction code of their own.

The essential rule is deceptively simple; namely, that lopsided eggs are laid in shallow nests.

There have been experiments to show that the eggs must be touching each other if they are to hatch at roughly the same time (and if they don't, the parent has a nasty dilemma: how to sit on the remaining eggs without sitting on the babies already hatched).

Uncommon Structures, Unconventional Builders

Above: Communal weaverbirds build enormous nests, like this one on a telephone pole in Namibia.

Right: A male village weaverbird (also called the spotted-back weaver) in full breeding plumage, perched on the twig where he has constructed a nest in the hope that some female will deem it a worthy nursery. If no mates accept his nest, he will tear it down and begin again.

Since eggs vary in shape, they differ in the distance they are likely to roll if jarred by a parent, predator, or a gust of wind. An egg that is much larger on one end than the other will roll in a very tight circle. One that is close to a regular oval shape is likely to roll farther and could more easily be nudged clear out of the nest. So for nearly all species of birds, the rule of thumb is the rule of egg: birds who lay lopsided eggs build shallow nests,

The Turkistan Remera builds nests with dried grass and feathers. Their hanging nests allow their offspring to be snug and warm.

If the resident is weatherproof,

the house needn't be.

Above: Some weaverbirds nest in groups—and burden the trees with the weight of their hanging thatch nests.

Left: Cliff Swallows build mud chambers in gourd shapes and attach them to cliff faces. The Cliff Swallows' main requirement seems to be the presence of a great many other swallows. In the nest itself, only a few feathers and a little grass are used as lining.

while birds who lay nearly symmetrical eggs build deeper nests to hold them.

Aside from keeping the eggs intact, this principle may be necessary for proper hatching. There have been experiments to show that the eggs must be touching each other if they are to hatch at roughly the same time (and if they don't, the parent has a nasty dilemma: how to sit on the remaining eggs without sitting on the babies already hatched). Some

The Least Bittern resides by shallow waters, where it grapples its nest to reed stalks and sometimes tangles the reed tops together to form a thatched roof over the nest.

Uncommon Structures, Unconventional Builders

The social weaverbird of southwest Africa
may have taught humans how to build
grass huts—leading the way with bird-sized
apartment houses in the trees, topped with
umbrella-shaped thatched roofs.

Left: The famous Baltimore Oriole, shown here in a drawing by Audubon, construct intricately woven hanging or semi-hanging nests out of dried grasses, yarn or any available fiber.

Opposite, top: English kestrels poach abandoned nests from other birds.

Opposite, bottom: The Ruby-throated Hummingbird is one of many species that use lichens to glue their nests onto tree branches. Other varieties of Hummingbird have other adhesives; the *Panterpe insignis* of Costa Rica sticks pieces of moss together with spider webs.

researchers have reported scratching sounds inside the eggs, infrequent at first and then speeding up as hatching time approaches. If the eggs are in contact, these sounds tend to synchronize, and the embryos develop toward emergence at the same pace. Although lopsided eggs would tend to remain in contact even in a shallow nest, a deeper cavity might be necessary to keep oval eggs in touch with each other.

Of course some species go well beyond the "rule of egg" in the sophistication of their nest building techniques. Of the 30,000 or so varieties of birds, here are a few of the more interesting architects and engineers.

The yellow-billed cuckoo makes a skimpy platform of twigs and straw so haphazardly thatched that the eggs occasionally plummet right through it.

Along the Amazon, the Japim weaves a sock-like nest out of tropical grasses and suspends the entire nest from a branch.

Uncommon Structures, Unconventional Builders

Virtually as inaccessible as they were two thousand years ago, the tombs of Petra rise out of the faces of rock cliffs in the arid Middle Eastern desert.

Petra

Match me such marvel, save in Eastern clime,

A rose-red city, half as old as time.

From the sonnet *Petra*
by John W. Burgon

In crudest terms, there are two ways to make a building: take pieces of stone or wood or something comparable and stack or fasten them together, or find a piece of rock somewhat larger than the finished structure is planned to be and hollow it out.

The hollowing method requires uncommon patience and boulders of such uncommon size that building sites are hard to find. A few times it has happened: once in the Arabian desert, at the city of Petra, and that site was so hard to find that for a thousand years the city was simply lost.

Midway between the Dead Sea and the Gulf of Aqaba in Jordan, Petra is surrounded by jagged rock cliffs. It is all but inaccessible and almost perfectly camouflaged; from the outside, nothing is visible but a forbidding range of craggy mountains. What it took to rediscover Petra was an adventurous Swiss aristocrat, disguised as a Bedouin, who was able to allay the suspicions of local Arabs and enter the hidden valley on the pretext of wishing to sacrifice a goat.

That was in 1812, when Johann Ludwig Burckhardt became the first European since the time of the Romans to see Petra and return to tell about it. Once a flourishing caravan center, the city had dwindled in influence, finally fell silent sometime before the seventh century, and was soon forgotten by the world—except

Petra lies in a great rift valley east of Wadi 'Araba in Jordan, just south of the Dead Sea.

Left: At the end of the cavernous path to Petra, visitors first glimpse the Khaznah Firaoun. The Khaznah, also called the Treasury of the Pharaohs, is one of the most famous structures at Petra.

Opposite, top: In the canyons and studding the cliff walls of Petra, tombs are everywhere.

Opposite, bottom: The most ambitious project in Petra is the 130-foot Dair, or Monastery, which was carved nearly 600 feet above the city on one of the peaks that ring the mile-square valley.

Under Nabatean rule, Petra had a flourishing spice trade. When trade routes shifted away from Petra, the city fell into decline.

for a small Arab tribe who believed in legends of great hidden wealth and guarded the secret and the treasure as theirs.

Burckhardt had not heard those legends. He was not looking for treasure or for Petra. He was merely roving the Middle East to learn Arab languages and customs so that he could later explore for the British African Association in North Africa. In Bedouin sheepskin and headdress, he tried to melt into the regional populace, but with imperfect success: three times desert sheikhs offered him their protection for a fee, then had him robbed and left in the wilderness. Each time he dragged himself back to Damascus—a little tougher, a little wiser.

On his fourth venture Burckhardt set a zigzag course for Cairo. Camped near the village of Aji at Wadi Musa—the Valley of Moses—he took advantage of his hard-won linguistics to learn that there were ruins nearby, including the tomb of Aaron, brother of Moses. Incurably curious, Burckhardt recruited a guide, somehow convincing the Arab of his fervent wish to sacrifice a goat at Aaron's tomb.

For Burckhardt, as for the visitor today, the route into Petra was a trip back through time. Down from the village he rode along the foot of steep, impenetrable slopes until his guide pointed to a narrow cleft where a stream entered the range of cliffs.

Uncommon Structures, Unconventional Builders

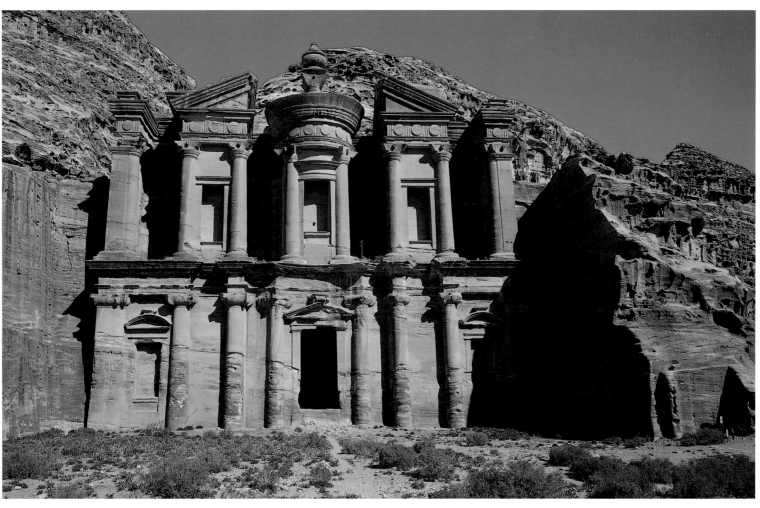

In this 1850 painting by David Roberts, the Dair is shown in detail. In the photograph on the preceding page, the weathering of the rock is evident.

Although excavation at Petra has been ongoing for almost one hundred years, only one percent of the city has been excavated, according to the research team at Brown University. Currently, the team is conducting the archaelogical excavations of the Great Temple.

If the guide had been suspicious at first, it was Burkhardt who grew wary as the gorge (called the Sik) deepened and narrowed. Between rock faces over 200 feet high, they picked their way through an eerie, boulder-strewn canyon over a mile long, until it squeezed to a width of less than twenty feet and seemed about to dead-end in darkness.

Then, through the narrow cleft ahead, Burkhardt glimpsed the sight that has stunned travelers ever since: a magnificent temple over a hundred feet high, its intricate façade of columns, capitals, statues, pediments, and entablatures carved out of the vertical rock face and almost perfectly preserved for close to 2,000 years. Brilliant in sunlight on the far wall of a transverse gorge, this massive mausoleum is called the Khaznah, "Treasury of the Pharoah." It is only the beginning.

Past the Khaznah, there are small family sepulchres and large royal tombs, all carefully hewn from the sandstone cliff faces. Where the gorge widens there is a theater large enough for two to three thousand spectators, its curving benches carved from rock strata of the canyon floor. Finally the great stone corridor opens onto a broad, enclosed valley nearly a mile square, and there is Petra.

Its most awesome wonders are natural. Incredible rock walls tower 600 feet above the valley. Surreal patterns of sandstone texture and strata trace great eddies and swirls that alter in hue and brilliance from hour to hour with the traversing sun. And color is everywhere: a bewildering interplay of red, orange, blue, gray, yellow, brown, purple, pink, white, and black. Reds predominate, but virtually no tint of the spectrum has gone unreflected by Petra's varying stone.

Once the capital and impregnable stronghold of the Nabataeans, Petra covered its slopes with temples, palaces, baths, private houses, and tombs—everywhere, tombs—interconnected by a paved

Uncommon Structures, Unconventional Builders

Once a flourishing caravan center, the city dwindled in influence, fell silent sometime before the seventh century, and was soon forgotten by the world—except for a small Arab tribe who believed in legends of great hidden wealth and guarded the secret and the treasure as theirs.

road that followed the meanderings of Wadi Musa. Stone staircases twist precariously up the slopes to the "high places" where stone altars received animal sacrifices to several generations of gods. There are at least eleven of these high places, most equipped with trenches to receive the blood and with water tanks, fireplaces, and stone seating for worshippers.

The active life of the city reached from about five centuries before Christ to five centuries after—roughly the span of Rome. The earliest known occupants were the Edomites referred to in the Old Testament, followed by the Nabataeans and the Romans. Finally there was a restoration of Arab rule. The ancient Greeks made at least one abortive military conquest of Petra, and contact between the two cultures continued. Caravans from India, Persia, Egypt, the interior of Arabia, and the maritime cities

Petra was conquered by Muslims in the 7th century and then briefly by the Crusaders in the 12th century.

The remains of an elaborate Byzantine church serve as a fragment of the once active city.

The façade of the Khaznah was used in the filming of the final scenes of the movie *Indiana Jones & the Last Crusade.*

Bedu guards at the Khaznah. The ancient city, which lay quiet and uninhabited for hundreds of years, is now a popular tourist destination. Booths selling tiny replicas of the solid rock buildings, carved olive wood mementos, and other souvenirs can be found near the excavated sites.

Once the capital and impregnable stronghold of the Nabataeans, Petra covered its slopes with temples, palaces, baths, private houses, and tombs— everywhere, tombs— interconnected by a paved road that followed the meanderings of Wadi Musa.

replacement of the stepped pinnacle by triangular pediments.

The Khaznah seems to mingle Roman and Hellenistic features, and a few structures—like the large mausoleum known as the Palace Tomb—are out-and-out Roman imitations, erected under the rule of colonial governors in the first two or three centuries A.D.

On every side are structures and public works whose execution is difficult to imagine. Four ancient obelisks stand twenty-three feet high on the south side of the wadi: the artistic and religious remains of a massive rock hill that was entirely carved away except for these four slender shafts, still rooted in mother rock. A huge tunnel, 30 feet high and 40 feet wide, bores 300 feet through solid stone just to serve as a storm sewer during infrequent heavy rains. Spring-fed the rest

of Syria and Palestine added to Petra's rich mixture of custom, culture, and architecture.

The Nabataeans soon developed a hybrid architectural style, distilled chiefly from Greek and Assyrian patterns and characterized by the use of stepped pinnacles. Later buildings evidence increasing Roman influence: larger, more imposing tomb facades, rows of columns, and

of the year, Petra is ribbed with channels carved in stone to conduct the water and with huge cisterns to store it. One of these channels carried water from a spring in the upper wadi through the Sik, where Moses is said to have drawn forth a spring by striking a rock.

At this date no one knows what metal tools were used to hollow out the large cubicle chambers of Petra's tombs and temples. Many of these rooms are designed to fit under black strata of rock, which are denser and harder than other layers of the sandstone, forming safe, durable ceilings. In the Khaznah the central chamber is forty feet square; two flanking chambers are each close to

thirty feet on a side; and all three have smaller rooms or niches carved into their walls.

Probably the most ambitious construction project in the history of Petra— and certainly one of the greatest whittling jobs in the history of anybody—was the conversion of a mountain top into the structure that Arabs called *ad-Dair* or *ad-Deir*, the Monastery (which it was not).

Carved from a massive peak of gray sandstone, the Dair stands 130 feet high and nearly 600 feet above the city. To hew it out of the mountain the Nabataeans had first to cut away a jutting cliff, leaving a stone mass over 150 feet wide and nearly that high from which to sculpt the

Hybrid forms of architecture help to date the buildings of Petra. Early structures of the Nabataeans combine Greek and Assyrian styles, while later buildings show increasing Roman influences.

According to the excavation team at Brown, the ancient city of Petra withstood two earthquakes, one in 363 AD that destroyed half of the city and another in 551 AD that nearly brought the city to ruin.

A Bedouin surveys the view of the valley
from the roof of the Dair.

Right: Nabataeans built tombs all along Wadi Musa's road, terracing them into the steep sandstone.

Below: The heights of the Nabataean hillside tombs are staggering—especially when seen from a helicopter.

Probably the most ambitious construction project in the history of Petra—and certainly one of the greatest whittling jobs in the history of anybody— was the conversion of a mountain top into the structure that Arabs called ad-Dair *or* ad-Deir, *the Monastery (which it was not).*

Uncommon Structures, Unconventional Builders

structure itself, and a broad area in front, which they leveled for an esplanade. Probably a temple rather than a tomb, the Dair is heavily Roman in design. Eight enormous columns support a pediment and eight more rise from there to frame a second story, centered by an immense urn atop a miniature temple complete with dome. All from one rock.

Niches and porches of the Dair, obviously intended for statuary, are now empty, their stone inhabitants carried off centuries ago. Throughout the city and its surrounding slopes, the tombs, temples, dwellings, and high places or worship have been stripped of their statues, icons, and urns, despite stern warnings on memorial inscriptions:

Petra is referred to in the Bible as Selah. There is a reference in II Kings 14:7 of the King James version where Amaziah, King of Judah, "took Selah by war."

Above: Although the city is far from the bustling metropolis it once was, it is still an evolving place. Dakhil Allah, the chosen leader of the Bedul Bedouin people, stands next to his own carving in the sandstone of Petra.

Right: Nabataean inscriptions on the Turkmaniyya tomb.

May Dushara and Manuthu and Quishah curse everyone who shall sell this tomb or buy it or frame for it any writ. The tomb and this its inscription are inviolable things, after the manner of the Nabataeans, for ever and ever.

What *did* prove inviolable, except for the gradual erosion of wind-driven sand and rain, are the essential structures of the city itself—too large to steal and too sturdy to smash because they were carved from solid petra, which, like Peter, means *rock*.

St. Peter healing the Lame Man, in a detail from the south transept portal of Beauvais Cathedral. Jean le Pot carved this scene out of wood in the 16th century.

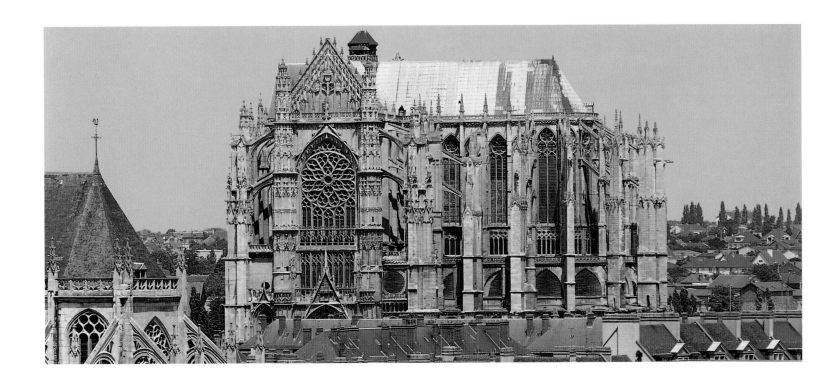

High Gothic

On Ascension Thursday the church fell down

For four years the tallest building in the world stood in the small French town of Beauvais, between Paris and Amiens. There, from 1569 to 1573, an unfinished cathedral stretched heavenward, its vault surpassing that of any other Gothic cathedral before or since, its spire rising higher than St. Peter's in Rome, higher by nearly twenty feet than the great pyramid of Cheops, higher than anything, anywhere, and higher than its structural underpinnings could long support. The Cathedral of St. Pierre at Beauvais is still standing. It lacks a spire; it lacks a nave. But it has a massive, magnificent choir and apse, an exceptional clock, and a strange, sad story of construction.

It was cathedral-building time in France when St. Pierre was begun.

During the twelfth and thirteenth centuries, a wave of church building unlike anything else in history spread through medieval Europe. France alone fostered at least 600 cathedrals and major churches and 500 abbeys between the years 1170 and 1270, and lowly was the town without a large church, or at least the beginnings of one.

Ahead lay the fourteenth, possibly the worst century yet counted by European man. Still further ahead, the rationalist eighteenth century would look back unsympathetically to the time of the great church building, would regard it as barbaric, and would slur it with the name "Gothic." Yet, it was the great synthesis of the Middle Ages, the period that tied each man, his society, and his religion together, and expressed these combined aspirations in stone pulled to unprecedented heights.

Midway in the period, in the year 1225, Milon de Nanteuil, Bishop of

A cathedral without a spire marks the site where, for four years, 400 years ago, the little French town of Beauvais could boast the tallest building on earth.

The Cathedral of Saint-Perre at Beauvais remains incomplete with a 13th-century choir and a 16th-century transept but it lacks a nave or a western frontispiece.

Even without its magnificent spire, the Beauvais cathedral is a staggeringly ornate building. Construction of the great cathedral began in 1225, and was taken up again in the 16th century under the direction of Martin Chamboges. The lavish carvings and tracery that was added to the exterior reflects an intervening shift to the *Flamboyant* phase of Gothic design.

Beauvais, ordered construction to begin on a new diocesan cathedral. In addition to its spiritual needs, both the pride and prosperity of Beauvais were at stake. Only thirty miles away, Amiens was erecting the nave and west façade of a cathedral that its builders boasted would be the tallest and most beautiful in France. Since major cathedrals drew thousands of pilgrims each year, economics alone might have been enough to prompt Beauvais to match its sister city and the hundreds of other French towns then busy with their buildings.

So construction began, but it did not, in the usual sequence of things, begin with the nave: that would unnecessarily delay the point, the towering spire that was planned to exceed the height of

France alone fostered at least 600 cathedrals and major churches and 500 abbeys between the years 1170 and 1270, and lowly was the town without a large church, or at least the beginnings of one.

Uncommon Structures, Unconventional Builders

Architectural historian Eugene
Viollet-le-Duc said of the cathedral,
"It is a Parthenon of French Gothic.
It has the misfortune, however, of
remaining unfinished and of not
being placed in a conserving population, like the ancient Greeks, able
to appreciate, respect an be proud
of the great efforts of human intelli-
gence."

When a church can't keep its stone intact,
what hope is there for windows? Modern
stained glass has replaced the original
panes, the last of which were lost to
bombings in 1940.

anything in Christendom. A small tenth-
century Romanesque church was left
standing on the site of the nave, while
work was started on the apse and choir
on which the spire would eventually rest.
Year by year the mighty structure rose
until, a half-century after its plan had
been announced, the apse and choir stood
complete, twenty feet higher than the
nave of Amiens and forty feet higher than
Chartres.

The ambulatory (the walkway around
the semicircular apse, from which seven
chapels radiated) was pulled from the
usual one-story height to a stupendous
three stories. The triforium and clerestory
above the inner aisle of the choir matched
this lofty level, and the choir vaults hovered an incredible 157 feet above the pavement, braced by two flights of flying
buttresses. Even without its spire St.
Pierre was the marvel of France.

In 1284 the vaults collapsed. Possibly
the foundation was insufficiently solid to
support such massive weight. More likely,
ambition had outrun engineering, and the
thrust and counter-thrust of uncommon-
ly high arches had been miscalculated. As
the choir was laboriously repaired, additional piering was added; however, before
the town and the diocese could muster
the courage and the funds to proceed

The triforium and clerestory above the inner aisle of the choir were three stories high, and the choir vaults hovered an incredible 157 feet above the pavement, braced by two flights of flying buttresses.

with further construction, a full two hundred years would pass.

A *very* full two hundred years. The calamitous fourteenth century devastated Europe with war, disease, civil disorder, and social dissolution. The Hundred Years' War bled the youth and wealth of France and England. The Black Death struck in the 1340's and left many regions with less than half their people still alive. What trade routes and pilgrimage routes had not been closed by war were choked off by pestilence. Towns languished; taxation soared; community and family life withered; the political and religious unity of the Holy Roman Empire began to disintegrate. There was no money for cathedrals, little will to build them, no guiding philosophy or purpose or design to shape them.

By the time work was resumed at Beauvais Renaissance architecture had produced the Rayonnant movement, which took the soaring verticals of the Gothic and experimented with stretching them as thinly and delicately as possible,

The choir, constructed between 1225 and 1272, was the tallest structure ever built in northern Europe and was the most ambitious cathedral project of the High Gothic era.

The massive stones of the unfinished charges of the vaults and towering uprights of the flying buttresses are a powerful expression of the ambitions of the builders of what was intended to be the greatest Gothic cathedral ever constructed.

Above: In its brief reach for glory, the spire of St. Pierre at Beauvais thrust 497 feet of stone and oak to an iron cross in the sky.

Opposite: Although St. Pierre's lacks a nave and a single spire, its lower towers reach high above the town of Beauvais. The church, with superb examples of Gothic arches, flying buttresses and decorative carving, is also known for its clock and magnificent choir.

The choir vaulting of St Pierre soars to an awesome 157 feet.

and the Flamboyant movement, named for the flame shapes it used in panes of glass and tracery. Both influences appeared in the final work on St. Pierre—Beauvais. Intricate Flamboyant facades grace the transepts, whose design was begun in 1500 by the renowned master Martin Chambiges and carried forward by Jean Vast and Michael Lalye. By 1548, both the north and south transepts were complete. There were still no plans for a nave, but discussions were already underway on design of the final triumph, the crossing tower.

In 1558, as the drum for Michelangelo's dome on St. Peter's was taking shape in Rome, Beauvais began its second bid for preeminence. Its new crossing spire would surpass St. Peter's 450-foot dome by a full forty feet. For eleven years the dream inched upward to its pinnacle: three stories of stone, one of oak, and an iron cross in the sky, 497 feet above the ground.

From the first, exultation in the town was mixed with concern. Suggestions were made for additional pierage and nave bays to stabilize support, and for replacing the heavy cross atop the spire with a lighter one. Finally, in the spring of 1573, masons were engaged to reinforce the tower's supporting columns. But it was too late.

On Ascension day the congregation attended services and began to leave the church. Only a few stragglers remained when an ominous scattering of dust and small stones filtered down from the upper reaches of the vault. Then, with a terrifying crack, the spire's supporting columns collapsed and, in five seconds, Beauvais had not the tallest but the shortest of all major Gothic cathedrals.

Gloom hung over the town. Even worse, jagged fragments of the spire hung precariously from a 200-foot stump and from edges of the damaged roof, threatening lethal showers of rubble into the square; and workmen refused the suicidal job of clearing the remains so that repairs could begin. The wreckage of Beauvais'

cathedral was destined to make only one man happy: a criminal awaiting execution in the town prison won his freedom by climbing to the spire and detaching the dangling remains.

Half a church stands today in Beauvais, where visitors can see an awesome, if incomplete, triumph of High Gothic. In 1929 one such visitor, author Sydney Clark, took particular note of an elaborate astronomical clock in the cathedral choir. Standing forty feet high, alive with 90,000 parts manipulating 52 dials, the great clock ticks off days, weeks, months, even the age of the Earth. Clark checked this age (5,730 years) through its biblical references and found that not only architecture but also time was out of joint at Beauvais. Its clock solemnly proclaimed the year to be A.D. 1726.

Despite its manifest difficulty, the building of the Beauvais Cathedral was not impossible. Some historians argue that the plans would have been perfectly practical if the architects had been blessed with better materials, such as good, hard stone, plus better financing and more capable help. The architects of St. Pierre-Beauvais had an additional problem: how to maintain control of a project that might require four or five hundred years to complete. Presumably the original designer would not have allowed construction of a spire that depended for support on a nonexistent nave. Cologne Cathedral, with proportions close to those of St. Pierre-Beauvais was left unfinished in the thirteenth century; but Cologne was completed in the nineteenth century, showing that St. Pierre could have been successfully engineered.

The transept was designed by the greatest master mason of French Late Gothic, Martin Chambiges, who directed the construction work until his death in 1532.

The spires at St. Pierre's famous cousin, the Chartres Cathedral, are sturdier, but less spectacular than the short-lived spire at Beauvais.

Roughly 1,500 miles of the Great Wall are still standing, admired now by tourists rather than generals.

The Great Wall of China

The raising of the wall

There is considerable evidence that the Chinese Emperor Ch'in Shih Huang-ti inadvertently toppled the Roman Empire. When his Great Wall discouraged the Huns from attacking China to the south, they turned west and invaded Europe, starting a six-century chain reaction that ultimately brought Attila to the gates of Rome. Had the Great Wall spanned the Urals instead of skirting the Mongolian plains, today's maps might show a different world entirely.

The fact that the wall is where it is proceeds first from an age-old Chinese preoccupation with walls and, second, from the febrile ambition of Ch'in Shih Huang-ti. Ch'in aspired to nothing less than immortality. He ordered the burning of all the books in China except scientific texts and his own memoirs. He anointed himself the first emperor and claimed divine stature. He banished his mother

(for a first emperor who is also a god, it is awkward to have a mother), and he searched ceaselessly for an elixir that would grant him eternal life. And, about 221 B.C., he launched the one enterprise that is still reaching for eternity: the Great Wall of China.

Ch'in had brought the warring states of China brutally to heel and was making plans for establishing a true empire, including uniform laws, currency, and weights and measures; a system of highways; and an end of feudal land holdings. To prevent barbarian incursions into his grand design, he ordered that various defensive walls along the northern boundary be connected into one immense fortification.

The emperor assigned the task to his commander-in-chief, Meng T'ien, and told him to hurry. Mindful of Ch'in's penchant for decapitating those who obeyed imperfectly, Meng hurried. He conscripted and

Stepped parapets high on the Great Wall at Huanghaucheng lead up to watchtowers every 100 to 400 yards.

The thickness of the Great Wall of China ranges from about four to 9 meters, which is about 15-30 feet and reaches 7.5 meters tall. The people in China call the wall "Wan-Li Qang-Qeng" which means "10,000 Li Wall" –"li" is about 5,000 kilometers.

Above: In this 1907 Herbert Ponting photo, the wall looked much like it does today, curving over high mountains. In some places, plants have grown up through the stones.

Left: In places, there are intricately carved figures in relief along the wall. These warriors can be found in the stones at Badaling.

Opposite: The basic design of the wall remained the same whether it cut through mountains or crossed the open plains. At completion, the wall had at least 10,000 watchtowers and more than enough masonry to build 2 million houses.

Laborers building the Great Wall were given mixed fermented pickles as part of their food ration.

Qin Shi Huang Di, who called himself the first emperor of China, ordered 700,000 workers and 300,000 soldiers to build a new structure that was strong and high to protect his people. The result was the Great Wall.

relocated almost an entire generation of Chinese manpower. To feed a construction crew numbering 1,500,000, he placed thirty-four supply dumps along the course of the wall. What provisions were not stolen in transit were lost to one of China's worst vermin plagues, but Meng pushed ahead.

By its description his construction site was a nightmare. The Chinese army, like all armies, wanted the high ground. Through mountainous terrain, Meng had his workmen chisel two parallel furrows,

Uncommon Structures, Unconventional Builders

The Great Wall of China: The raising of the wall

万里長城

室家墮茨圖

Construction of the Great Wall figures largely in Chinese art. The wall took more than ten years to build and 1,500,000 laborers worked on it. If a worker was unfortunate enough to die during work on the wall, the corpse was buried in the wall itself.

The Great Wall is made up of two side walls, 25 to 30 feet apart, made of granite blocks topped with fire-clay bricks.

often out of solid rock. From these trenches, twenty-five to thirty feet apart, they built the main walls on massive footers of granite blocks, piled four to five feet wide and from six to twelve feet high. These were topped by outer walls of large fired-clay bricks, two feet thick, in what may have been the first large-scale use of this building material.

At their tops the outer walls were about twenty feet apart and from eighteen to thirty feet high. Between them, workmen dumped earth, rock, and clay, trampled this filler with their feet, and probably tamped it with logs. By most accounts, the core material also contained the thousands of laborers who died during construction. Finally a brick platform was built over the top to provide an elevated road.

Flanking this narrow cartway were crenelated parapets that doubled as guardrails and as protection for defending archers and spearmen. And rising another twenty feet from the top of the wall were uncounted watchtowers—*literally* uncounted, but there were at least 10,000 towers on the wall and perhaps as many as 25,000— one every 100 to 400 yards.

Uncommon Structures, Unconventional Builders

When its course left the mountains for the open plains, the wall retained its basic design, but construction methods changed. Workmen dug the same parallel trenches but saved the dirt, which contained a high percentage of a yellowish, calcareous substance called loess. Meng had wooden forms built to frame the outer walls. When the loess was rammed into these forms and a little water added, it stiffened and baked in the sun to form a hard, dry wall. Some of this loess surface was faced with brick, but much of it was left bare, to be gradually worn away by centuries of wind from the deserts.

From these trenches, 25 to 30 feet apart, they built the main walls on massive footers of granite blocks, piled 4 to 5 feet wide and from 6 to 12 feet high. These were topped by outer walls of large fired-clay bricks, 2 feet thick, in what may have been the first large-scale use of this building material.

All told Ch'in added at least 1,200 miles of wall to existing structures. With later additions in the Ming dynasty, the Great Wall reached a length of 2,500 zigzag miles, spanning a 1,500-mile frontier. One historian who visited the wall around 1800 calculated that it contained more masonry than all of the nearly two

The square towers, which dot the wall every 100 to 400 yards, date to about 1368, the beginning of the Ming Dynasty, when the wall was repaired.

The wall was made in layers—brick making up the walkways, towers and parapets on top of a thick foundation of granite blocks.

Part of a rammed-earth wall erected during the Han Dynasty snakes across brown soil in this sunset view. The Han wall and fragments of the Qin wall were eventually supplanted by the Great Wall.

The Great Wall is totally abandoned in some places, left for nature to take over. In the mountains, the foundations were granite—but in the plains, they were made of a sun baked brick that has eroded more quickly.

Uncommon Structures, Unconventional Builders

With later additions in the Ming dynasty, the Great Wall reached a length of 2,500 zigzag miles, spanning a 1,500-mile frontier. One historian who visited the wall around 1800 calculated that it contained more masonry than all of the nearly two million houses in England and Scotland. Ch'in's portion was completed in an incredible ten years, at a cost in human life that has been estimated at more than one million men.

million houses in England and Scotland. Ch'in's portion was completed in an incredible ten years, at a cost in human life that has been estimated at more than a million men.

Ch'in had little use for scholars, and scholars have had little use for him. One of the kindest things ever said about him was that he ruined one generation of Chinese to save many generations. His wall, though it didn't prevent invasions, certainly delayed them and reduced their number. And if Ch'in failed to find the elixir to cheat death, he succeeded in cheating obscurity. He is remembered as the man who burned the books and is memorialized by the hundreds of miles that remain of the Great Wall. He also had a country named after him—China.

The barrier walls at Jinshanling show little wear after thousands of years of wind and weather.

Opposite: Chinese and foreigners walk along the Great Wall at Badaling early in the morning of January 1, 2000. Hundreds gathered to tour the Great Wall, which was lit up for the occasion, and watch performances as they waited for the year 2000 to arrive.

Neil Armstrong reported from orbit that he could only see two man-made structures from space: The Great Wall of China and the dike system of Holland.

Igloos

How to build a northern hemisphere

There is no practical way to take an accurate census of Eskimos, but best estimates see roughly 50,000 of them scattered over a geographic region 3,200 miles long—all of it cold.

The arrival first of traders, then prospectors, military contingents, research stations, and finally oil drillers and pipeline companies could easily cause Eskimo culture to disappear before it is well understood. In many Alaskan and Canadian Eskimo communities, electrical wiring, wooden frame buildings, and snowmobiles have replaced whale oil, igloos, and dogsleds. With a radio in each house and a stereo in the village co-op, young people take up the six-string guitar and ignore the Eskimo skin drum. The igloo becomes not a mainstay residence but a field expedient for camping. Boy Scouts build them, and hunters, and the Canadian Air Force on bivouac.

For the Eskimo not yet touched by this creeping encroachment, life is a harsher, more difficult, though possibly more coherent set of realities. He speaks a complex language, masters an elaborate system of religious belief in which every object is possessed by a spirit, and lives in a primitive commune which intermingles monogamy, polygamy, and polyandry. His architecture, since it is a pure matter of survival, is less complicated.

The Eskimo family has a winter home and a summer home, each made of the building materials that nature provides in season and ready to use. The dwelling for summer may be a tent made of caribou hide, which offers the only physical property required: waterproofing. In winter, depending on the locale, a home may consist of mud and stone; but in regions of unrelenting cold and endless snow it is the snow itself that furnishes thermal insulation and shelter.

Above: Modern tools have made igloo construction a little quicker, but the process has remained essentially the same for hundreds of years.

Opposite: Before returning to his camp, Lypa Pitsiulak, an Inuit living on an outpost camp in the Opingivik area of Nunavut, Canada, covers the opening of an "iglu," Inuktitut for igloo. Behind him is an "inuksuit," a human shape used by the Inuit as a landmark.

By building a series of intersecting domes and opening up the common segments, an igloo intended for one or two families can be enlarged into a small circular community capable of housing four or five families.

For the Eskimo not yet touched by this creeping encroachment, life is a harsher, more difficult, though possibly more coherent set of realities.

For economy, simplicity, and speed of construction, the igloo may never be matched. A man and his wife can build one in an hour, complete with skylight, using no tools other than a simple bone knife. But there is a knack to it.

Not far from prime seal hunting grounds, the Eskimo picks a clear, level spot on the frozen snow and lays out a circle ten to fifteen feet in diameter. Then he carves blocks of snow from within the circle and lays them around the perimeter in an upward spiral, thus excavating and raising walls at the same time. He piles these snow blocks in narrowing circles until he has formed a hemisphere, with an opening left at ground level for access and another near the top.

Into the higher of the two openings he puts a piece of clear ice to serve as a window or skylight. This last piece is a little harder to find, so it's not uncommon for the Eskimo family to carry around a choice piece of ice for several weeks before igloo-building time.

Once the snow-block beehive is up, the Eskimo packs every crack with loose snow. Then his wife takes over.

She crawls in through the door opening—which is designed to be just large enough for her but too narrow for, say, a polar bear—and lights a blubber lamp inside. Feeding the lamp constantly with whale oil or seal blubber, she gets a small fire started, then seals the entrance and builds the fire higher.

It isn't long until the entire igloo softens and begins to melt. Because of its hemispherical contour, the water runs down the walls instead of dripping. When the entire structure is damp, she opens the entrance, and the cold blast of Arctic air freezes the walls and welds the structure so solidly that nothing short of the spring thaw can damage it.

Inside, the snow igloo is not only snug but surprisingly well lighted. Aside from the oil lamp or cooking fire and the window of ice, the snow blocks themselves are translucent enough to admit a little sunlight, or even bright moonlight.

Inuit igloo communities were busy places—full of dogs, children and the work of survival in a harsh environment.

Uncommon Structures, Unconventional Builders

Furniture is simple: a three-foot-high snow shelf at the rear, to serve as a family bed. Remaining floor space is left for cooking, eating, and storage, although most of the family provisions are kept in a tunnel—usually about fifteen feet long— built out from the entrance.

In the icebound northernmost regions this building technique has remained unchanged for hundreds of years. The only recent innovation is a questionable one: some Eskimo families have now acquired iron stoves through their limited trading, and the overheated result is often a slowly

To seal the walls of an igloo, a layer of ice is created. After the doorway ice block has been sealed, a lamp is kindled inside. The heated air, having no exit, begins to melt the face of the snow blocks, which drips down the sides of the snow walls. The doorway is then opened to allow outside air to enter the igloo, freezing the melting snow on the walls into ice.

3-year-old Gordon Pitsiulak climbs out of an igloo on an ice floe on Baffin Island, Canada.

For economy, simplicity, and speed of construction, the igloo may never be matched. A man and woman can build one in an hour, complete with skylight, using no tools other than a simple bone knife. But there is a knack to it.

melting igloo that needs constant bolstering and patching.

Notes for the backyard igloo-builder

Temperate-zone urban or suburban igloos require considerable compromise. The structure an Eskimo would be likely to build in Westchester or Evanston would involve several tons of wood, sod, and stone—and the neighborhood children would never accept it as an igloo. Best bet

Igloos have become vogue vacation spots; one Greenland hotel is made entirely of ice and snow. The Hotel Igdlo Village in Kangerlussuaq, Greenland boasts five igloos with room for one or two people each, an ice bar and daily excursions complete with a cold drink and a lesson on the nature of ice.

European explorers at the turn of the 20th century were fascinated by Inuit culture, and took a number of photographs of their homes and habits. The family pictured here lived on the banks of the River Glyde on Baffin Island.

Opposite: Roald Amundsen met and photographed many Inuit families on his expedition to the North Pole (which he reached in 1909).

is to ignore the lack of north latitude and build an Arctic snow-block igloo anyway.

Snow blocks may be molded in a wooden box or cut from large, cold-rolled snowballs. For the window a piece of clear polyethylene film can simulate lake ice. You can't use the melt-freeze solidifying process that Eskimos use in subzero temperatures, but you can certainly use a sprinkling can. On a cold day, a little water on the outside of the igloo will do

Uncommon Structures, Unconventional Builders

Roald Amundsen spent years exploring the Arctic, where he met this Inuit couple between 1903 and 1907.

the job, at least well enough for a neighborhood where there is no polar bear to attempt a forced entry.

Variations on the theme
Further south, where hard-frozen snow is less abundant and vegetation more plentiful, the construction technology changes. Although snow is often used as an exterior insulation layer, the basic structural materials are various

Inuit, the native people of the Arctic region, built igloos that could hold more than 60 people with room for drumming, dancing and singing performances during long winter nights.

combinations of logs, twigs, driftwood, mud, moss, and stone.

In some regions the Eskimos bend long willow branches into a series of arches to form a skeletal dome, then add layers of moss, dirt, and sod. For a window they use an animal skin with a peephole in it. If there's a bearded seal in the vicinity, his intestine will become the windowpane: Eskimos have found it to be not only weatherproof but also nearly transparent.

Air conditioning in this type of structure is crude but effective. A hole left in the sod and moss at top center serves as a smoke ventilator. An animal skin over the igloo entrance can be partly or fully opened for fresh air intake, and a vent hole in the window can be closed by stuffing a whisk of hay through it.

The sod-wood-stone igloo often becomes a permanent home and receives

Uncommon Structures, Unconventional Builders

such refinements as a stone fireplace under the smoke hole and raised stone or earthen beds covered with animal skins. Tanning and curing of these skins—not only for bedspreads but also for clothes, shoes, windows, doors, and tent walls—could be a problem. The Eskimos have a workable solution: their women simply chew the skins, hour after hour, until they are soft and flexible.

Nature and nurture
In each climatic region, the Eskimo takes the natural building products that are close at hand and improves them a little, but without interfering with their best natural properties. The melt-freeze treat- ment of snow blocks hardens their crust for structural strength but does little to impair their insulating efficiency. Mama Eskimo's endless chewing on caribou hide and sealskin softens it but leaves it water- proof, not only for building but for clothing and boots. That it finally leaves her toothless is one of the penalties of Arctic life which fewer and fewer Eskimo women will be willing to pay.

Not all arctic dwellings are made of ice— some are made of thatch or pelts, like this one in Greenland (photographed in 1930 by the British Arctic Air Route Expedition).

On leaving an igloo, Inuit have been known to take advantage of the Igloo's privacy. Some lore suggests that after making a final visit to the igloo that would soon melt with spring's warmer temperatures, Inuit would allow their Huskies, dogs known as voracious excrement eaters, to enjoy the steamy treat.

Modern adventurers use igloos for shelter on multi-day ski trips.

Viking ships were renowned for their design. This example, the Gokstad ship, now in a museum in Oslo, shows extraordinary workmanship that has lasted for centuries.

Heorot

Beowulf at home

Thirty men lie sleeping on hard wooden benches, in alcoves, and on the drafty floor of a great timbered hall: a combination fortress, castle, barracks, and rathskeller. Fires burn at two hearths, but the occupants—soldiers with shields and swords at their sides—are wrapped in animal skins against the chill northern night. There is a sudden crash. An enormous fist fells the split-log door with one blow, ripping it from its iron bolts and hinges. Every eye opens but the bodies are motionless, waiting, as a monstrous manlike creature, still dripping mudwater and trailing moss from the moors, advances on the prone warriors, his eyes burning. One man is seized and, as his comrades watch, frozen with terror, is devoured on the spot. The next man in the monster's path is not terrified: he has watched the attack and the gory feast as a study in fighting style, and he is ready when the intruder—

Grendel—reaches for him and finds to his consternation that he has chosen the wrong victim. The great beast Grendel is suddenly in the grip of Beowulf, and he has eaten his last warrior.

The epic battle between Beowulf and Grendel has thrilled English-speaking people for a thousand years. It cost Grendel an arm, and the beast fled to die on the moors. It incited a retaliatory raid by Grendel's mother, a giant water hag who turned out to be even meaner than her son. To end *that* menace Beowulf had to fight mother alone in her cave, so deep under a lake that it took him an entire day of diving straight down to reach it. And somewhere, buried beneath layers of magic and marvel, lies the memory of a great mead hall where the first battle took place.

This was Heorot, site of the great English folk epic. It was not in England, but it did exist—in an age when real war-

This Viking stone relief of the Legend of Valhalla was created in the 9th century, and is now housed in the historical museum in Stockholm.

Around 860, Norsemen went to the Faeroe Islands northwest of Scotland. A decade later, the first of some 12,000 Vikings landed in Iceland.

HWÆT WE GARDE

na ingear dagum. þeod cyninga
þrym ge frunon huða æþelingas elle
fremedon. oft scyld scefing sceaþe
þreatum moneʒū mæʒþum meodo setl
of teah egsode eorl syððan ærest wear
fea sceaft funden he þæs frofre geba
weox under wolcnum weorð myndum þah
oð þ him æʒhwylc þara ymb sittendra
ofer hron rade hyran scolde gomban
gyldan þ wæs god cyning. ðæm eafera wæs
æfter cenned geong in geardum þone god
sende folce tofrofre fyren ðearfe on
geat þ hie ær drugon aldor ... ase lange
hwile him þæs lif frea wuldres weal dend
worold are for geaf beowulf wæs bremen
blæd wide sprang scyldes eafera scede
landum in. Swa sceal .... ʒodum
ʒe wyrcean fromum feoh ... ... um oð gewriten

Opposite: Beowulf, one of the greatest myths in history, has been interpreted in many languages. The most famous version is the poem that begins with this page.

Every hero must have an enemy—Beowulf took on the fearsome Grendel after the vicious creature attacked Beowulf's homeland.

Woden, according to Norse mythology, was a god of warriors. It was his role to select, from those fighting in battle, which were to be victorious and which were to be slain. The fallen warriors were taken to Valhalla where he presided.

riors fought real monsters (wolves, bears, and boars, if not fabulous beasts). They most certainly fought each other, and, in calmer times, they feasted and drank mead in real mead halls. Aside from Valhalla, and, assuming that Camelot was a castle rather than a hall, Heorot is the most renowned example of the mead halls that dotted England and northern Europe in premedieval times.

The fighters and feasters who inhabited these structures were the very early Vikings. Their way of building, like their way of life,

was fairly crude, though often ingenious. Since their only architects were shipwrights, and since they had little use for stone ships, building design and construction centered around the working of wood and seldom strayed far from the configurations of boats: right-side-up at sea, upside-down on land. Norse builders erected wooden halls, houses, even castles, in places where stone was plentiful and timber scarce. So mead halls were timber and, like the American log cabin, were developed in Scandinavia and later exported.

Beowulf, a member of the Geats, heard of Grendel's attack on his homeland. Hygelac, king of the Geats, selected 15 of the bravest men to help the Danes.

Uncommon Structures, Unconventional Builders

Around 1000, Leif Ericson sailed with 35 men to Newfoundland, known in the Norse sagas as Vinland. Thorfinn Karlsefni followed with his wife, Gudrid, and a party of 160 men and women in three ships, staying three years. Thorfinn and Gudrid's son, Snorri, is said to be the first European born in America.

Archaeologists are still uncovering Viking artifacts. These three keys to Viking buildings were found at the Coppergate site in York, England.

The roots of *Beowulf*, like those of *Hamlet*, are in Denmark. Beowulf and his warriors were Geatas, whose home was southern Sweden early in the sixth century. They went to the aid of King Hrothgar at Heorot, probably on the island of Zealand off the Danish coast, and the story of their exploits was carried to England by Norse invaders who settled there in the seventh century. Handed down by word-of-mouth from one generation to the next, growing with each retelling, the legend was finally written down no earlier than the eighth century and possibly as late as the tenth, when the

Hrothgar's Heorot is likely to have been located on the island of Sjaelland near the present day city of Roskilde.

surviving manuscript was prepared by English monks.

By that time, the actual events that had inspired the poem had been magnified to mythic proportions. To begin a description of Heorot is to start with structural specifications screened through the beautiful but baffling overlays of poetry and lore as they appear in the poem itself:

> To Hrothgar in time came triumph in
> battle,
> The glory of the sword, and his
> friendly kinsmen
> Flocked to serve him till the band of
> them was great,
> A host of eager retainers. And his mind
> Stirred him to command a hall to be
> built,
> A huger mead-house[1] to be made and
> raised
> Than any ever known to the children
> of men,

---

1. Main functions of the mead house or hall were sleeping, eating, and carousing; and mead was the ever-flowing beverage: a fermented mix of honey, malt, water, and yeast.

The Scylding line is known through Scandinavian and Anglo-Saxon sources; the Anglo-Saxon king Cnut (1016-1042, a period coincident with the composition of the *Beowulf* manuscript) is known to have descended from this line.

Where he under its roof to young and
    old
Would distribute such gifts as God
    gave him,
Everything but the lands and lives of
    his people.
Not few, we are told, were the tribes
    who then

Were summoned to the work far
    throughout this world,
To adorn that dwelling place. And so
    in due time,
Quickly as men laboured, it was all
    prepared,
Most massive of halls; and he called it
    Heorot....The hall towered up
Clifflife, broad-gabled.[2]...

(Before Beowulf's arrival,)
He (Grendel) went then to visit at the
    full of night
That lofty hall, to see how the Danes
Fared as they lay at the end of their
    carousing.

2. Gables of Heorot were not only broad but high, and were adorned by stag antlers.

When Viking shipbuilders turned their talents to building houses and halls, the same timbers turned up with the same joinery that held their boats together and, as seen here, even some of the same curving contours. This reconstruction of a craftsman's home is in a replica of a Viking trading village in New Hedeby, Denmark.

Vikings filled their graveyards, such as this one in Denmark, with symbolic boats of stone to accompany their dead warriors on explorations in the next life.

According to an old Icelandic poem, Woden was hanged on a tree for nine days and nights, and was voluntarily pierced with a spear. As a result he learned wisdom—the secret of runes.

Within it he found the band of warriors
Sleeping after the feast....
...The creature was like a pestilence
    raging and ravenous, quick at his
    task,
Savage and unsparing, seizing thirty
    Soldiers from their beds....
Then it wasn't rare for a man
    elsewhere
At a greater distance to look for t
    his rest,
For a bed in the outbuildings[3]
    ...till empty and unvisited
Stood the best of halls.

(Beowulf's)...men hastened,
They marched together until they
    could see
A timbered building resplendent
    with gold:
The most illustrious hall under
    heaven
For men to move in, and home of
    a king:
Like a lantern illuminating many
    lands.[4]
Here the soldier showed them the
    dazzling
Hall of the heroes....

3. Outbuildings usually included cookhouse and storehouses, sometimes a buttery and a smithy, and separate sleeping quarters for women and children.

4. According to legend, Heorot's roof glinted with gold trim, reflecting flashes of sunlight visible far out to sea.

Uncommon Structures, Unconventional Builders

Grendel came stalking; he brought
  God's wrath.
His sin and violence thought to
  ensnare
One of our kind in that hall of
  halls.
He moved through the night till with
  perfect clearness
He could see the banquet-building,
  treasure-home of men
Gold panelled and glittering....
  ...The door, firm iron-bound,
Flew at once ajar when he breached it
  with his fists:
The hall's mouth he tore wide open,
  enraged
And possessed by his evil. Quickly
  after this
The fiend stepped over the stone-
  patterned[5] floor.
(As Beowulf battled Grendel inside
  the hall:)
...Both hosts of the hall
Were in rage, in ferocity. The building
  reverberated.
It was more than a marvel that that
  wine-house
Stood up to the battle-darers, that the
  splendid walls
Didn't fall to the ground; but it had
  the solidity.
It was cleverly compacted both inside
  and out
With its bands of iron.[6] Many were
  the mead-benches
Sumptuous with gold that
  were wrenched there from the
  floor,
Beside the antagonists in their epic
  fury.
—A thing undreamed of by Scylding
  wisdom
That any man could ever, in any
  manner,

Shatter it as it stood stately and horn-
  spanned[7]
By sleights disrupt it: till fire's[8]
  embrace
Should become its furnace.

(When the fight was over,)
Severely shattered was the shining
  building,
For all its inward iron-band-buttress-
  ing;
Its door hinges were sprung; only roof
  escaped
Without any damage when the
  slaughter-guilty
Deed-stained monster turned away in
  flight
Despairing of life.[9]

When a story takes from two hundred to
four hundred years to ripen before anyone
writes it down, the details of its architecture are necessarily hazy. Archaeology can
often help but, in this case, not a great
deal. Thousands of graves have been
excavated throughout Scandinavia and
the Norse settlements in Britain. Many
Viking ships have been raised and some
reconstructed. But few buildings remain,
and most of the survivors are houses:
royal halls had a way of getting burned
down after battles.

   However, there are clues in the fasci-
nating kinship between the longhouse or
"fire hall"—the typical farm dwelling
throughout the Viking age—and the other
basic unit of Norse construction, the war-
ship. One conjecture has it that primitive
dwellings in northern European coastal
regions often began with an overturned
boat as a roof, supported on timber posts.
Side walls of the early longhouses show a
marked curvature—outward at the center
and inward toward the gabled ends—in a
boat shape that would mate with such
a roof; there appears to be no other

*Heorot is "a great mead-
building that the children of
men should hear of forever."
Hrothgar builds Heorot as a
home for his warriors; this
action is similar to creating a
civilization out of chaos.
Heorot attracts Grendel, who
attacks the hall for 12 years,
until Beowulf slays him.*

---

5. It's not clear whether the floors of Heorot were
   actually stone or wood. References to mead-
   benches being bolted to the floor seem to favor
   wood in a parquet "stone patterned" arrangement.
6. Viking timber structures used iron-band joinery, a
   technique borrowed from their famed art of shipbuilding. A buried Viking ship discovered at
   Sutton Hoo on the southeastern English coast
   showed remnants of these ironbound joints.

---

7. A reference to the fact that Heorot's roof was
   adorned with stag horns. Heorot means *hart* or
   stag.
8. Heorot was finally burned in the Danish
   Heathobard vendetta.
9. *Beowulf* excerpts from verse translation by Edwin
   Morgan originally published by The University of
   California Press; reprinted by permission of The
   Regents of the University of California.

Much of the English language
comes from Norse: born, die, sister,
window, plus the days Tuesday,
Wednesday, Thursday and Friday,
named for the gods Tyr, Odin, Thor
and Frey.

Exiled from Iceland for manslaughter,
Eric the Red spent three years exploring
Greenland's coastline. On his return
home, he convinced settlers to join him
in colonizing the head of Eiriks Fjord.
They founded Brattahlid, where the stone
foundations of an early 14th century
church still stands.

Fearsome warriors and restless predators, the Vikings raided, plundered, and often colonized the English, Scottish and European coastlines from 800 to 1100.

Below: A close-up view of wall shingles of a reconstructed Viking house reveals that the shingles were hand-hewn with an adze.

The Viking Age started around 793, when sailing Norsemen started capitalizing on the poorly defended coasts of Britain, Ireland, and mainland Europe with a raid on England's Lindisfarne monastery.

explanation for the curves. Whether or not an actual boat bottom ever topped a house, there is little question that home building in great seafaring cultures of the heroic age was dominated by Norse shipwrights whose talents for timberworking and joinery were the marvel of Europe.

Curiously, one point where these two trends most clearly converge is the remnant of Trelleborg fortress, on the same island once occupied by Heorot and defended by Beowulf. Trelleborg guarded the west Zealand coast in the tenth century, at the height of Danish sea power under Cnut the Great, four centuries after the events of *Beowulf*. If a longhouse is reminiscent of an overturned Viking ship, this fortification can be seen as a capsized fleet. Its symmetrical layout enclosed sixteen garrison buildings arranged in four squares, all within a circular rampart, and fifteen slightly smaller houses outside. Curved sidewalls and straight end walls gave each house at least a crude resemblance to a truncated ship turned upside down, its ridgepole arched like a keel. It may have been more than coincidental that each building (roughly a hundred feet long) had enough space to garrison one shipload of fighting men.

Like the smaller longhouses and the larger mead hall, the Trelleborg style of barracks building had a large stone hearth near the center, its fire vented through the roof, and raised sleeping platforms along each side wall. Apparently these barracks—and others defending headlands and key coastal regions throughout Scandinavia—were more Spartan versions of the great mead halls: more military and somewhat less festive, although it is true that excavation of grave sites in the Trelleborg camp has turned up the remains of a number of women.

As to size, there is nothing left of Heorot to measure. From this point in history its proportions are no easier to assess than those of Camelot, which

Woden's hall was described as having a framework of spears and a roof of shields, form following function to the extent that Valhalla was the place where fallen warriors were sent after death, fully equipped with spears and shields.

Uncommon Structures, Unconventional Builders

THE PROW.

FIGURE SHOWING CONSTRUCTION, AND VARIOUS IMPLEMENTS AND ORNAMENTS FOUND IN THE MOUND.

SUPPOSED ORIGINAL APPEARANCE OF THE VESSEL.

REMAINS OF THE VESSEL AS FOUND IN THE "KING'S HILL."

A VIKING'S SHIP, DISCOVERED NEAR SANDEFJORD, NORWAY.—[SEE PAGE 518.]

supposedly flourished in England at roughly the same time and has since undergone a similar escalation into legend.

Both, it may be assumed, were less ambitious in specifications than the great granddaddy of all mead halls, Valhalla, the home of Woden in Teutonic myth or Odin in the northlands. Valhalla had 540 doors, each wide enough to admit 800 soldiers abreast. Allowing three feet for each warrior with his shield, and assuming that walls occupied three times the width of doors, Woden's mead hall was 5,184,000 feet in circumference, roughly the size of Pennsylvania!

One might expect Valhalla, like Heorot, to be a timber structure, since in the traditions of Teutonic myth and Norse poetry the universe was supported by a gigantic tree. But in fact Woden's hall was described as having a framework of spears and a roof

Harper's Weekly ran drawings of Viking tools and ship details when the remains of a ship were found in Norway in 1880.

of shields, form following function to the extent that Valhalla was the place where fallen warriors were sent after death, fully equipped with spears and shields.

In Iceland, where neither spears nor timbers were plentiful, Norsemen overcame their predilection for timber in boatbuilding configurations and instead erected stone walls and turf roofing. Viking settlers in the Orkneys and Shetlands used alternating layers of stone and earth or turf. Norwegians under Eirik, or Erik , left remnants of similar structures in Greenland. Had they left the same traces in "Vinland," they might have prevented a good deal of argument over who discovered America.

Woden was an important Anglo-Saxon god, associated with the Scandinavian Odin, and Wotan from Wagner's *Ring*. He was also known as Grim in the Saxon kingdoms, and gave his name to Wednesday. He was identified with the Roman god Mercury (Wednesday—Mercredi in French and Mercoledi in Italian.)

Heorot: Beowulf at home　　　　131

The Trouble with Cabins and Cottages

Early American alternatives to the log cabin and the White House

The great houses of early America and the architectural influences they embody have been widely noted, thoroughly documented, and often restored or replicated. Humbler dwellings, quite properly, have received less frequent and less lavish attention. Not so properly, the cabins and cottages in which the overwhelming majority of early Americans were raised have been somewhat garbled in popular histories. We can, however, explore a few of the curious twists through which the American cabin and cottage have come.

In the early 1840's, with the help of William Henry Harrison's "log cabin and hard cider" campaign, the log cabin clinched its place as a symbol of America and all but dislodged other early dwellings from the national memory. (A superb documentation of how this happened may be found in *The Log Cabin Myth* by Harold

Shurtleff, introduced and edited by Samuel Eliot Morison and published by Harvard University Press in 1939.)

Architects, however, were not leaving modest dwellings and humble origins entirely to the politicians. While Harrison and Tyler stumped the country, visiting the party faithful in hastily built log-cabin headquarters, horticulturist and "rural architect" Andrew Jackson Downing was working on a book to be called *Cottage Residences*. And architect John Bullock, editor of a textbook series on the rudiments of architecture, was planning to publish *The American Cottage Builder*. Both books appeared between the two great log-cabin campaigns: those of Harrison and Lincoln.

As for the log cabin as an architectural genre, it is roughly as American as apple pie, which was invented by the Romans. The familiar sight of a crude but sturdy dwelling of round logs—notched, crossed

Above: The Pilgrim fort at New Plymouth was one of the first European attempts at cabin-building in the New World.

Opposite: A reconstruction of a bark-covered, wigwam style shelter, of the type built by the early settlers, sits at a pioneer village in Plymouth, Massachusetts.

The log cabin emerged as an all-American symbol with the help of two famous political campaigns. It was virtually ignored by settlers in the original colonies.

But the countryside was just as the emigrants had left it: the houses were framed with timbers and walled with clapboard or wattle and daub, and those were the only houses that the colonists knew how to build, or would want to build.

at the ends, protruding at each corner, chinked with clay, and topped by a roof of split oak shingles—was not at all a familiar sight to the first settlers who colonized this country, a century of grade-school history textbooks notwithstanding.

Jamestown was settled by English small-town and country folk in 1607. Back home James I was feuding with the House of Commons over union with Scotland. Oliver Cromwell was flunking arithmetic at the Huntingdon free school. Scholars at Oxford were shaking their heads over odd new theories from someone named Kepler in Germany and Galileo Galilei in Italy. Londoners were still buzzing about William Shakespeare's new play, *Antony and Cleopatra*. But the countryside was just as the emigrants had left it: the houses were framed with timbers and walled with clapboard or wattle and daub, and those were the only houses that the colonists knew how to build, or would want to build.

Above: Log cabins are still a viable option for people throughout North America—this one houses the Crow Creek Blacksmith's Shop in Alaska.

Opposite, top: A wizened settler guards his cabin in 1885, on the banks of the North Thompson River in British Columbia.

Opposite, bottom: The United States' most famous log cabin: Abraham Lincoln was born here in 1809. The cabin stands on the Kentucky farm that belonged to Lincoln's father and still exists, housed within a museum building.

Uncommon Structures, Unconventional Builders

The Trouble with Cabins and Cottages: Early American alternatives...

This traditional cabin near Front Royal, Virginia, shows a slightly more refined design, with a stone foundation and hewn corners. By 1700, the log cabin was a favorite first home for settlers who had the Northern European knack of notching logs for tight fits and plumb corners.

Even earlier roots have been found in Stone-Age European lake villages, where some houses used crossed logs, overlapping at the corners.

At first they erected temporary shelters, notably wigwams, which they also knew how to build because the basic bentwood construction technique was similar to that of the English garden arbor. But soon the sawyers and carpenters were at work, riving clapboard; hewing timbers for joists, studs, corner posts, and rafters; and neatly mortising corner joints for permanent frame houses. It would be thirty-one years before an English colonist would see his first log cabin, and a good deal longer before any of them would see fit to build one.

When the log cabin did arrive, it came from Sweden. Apparently it was the Scandinavians who invented the architectural link to Lincoln; but by the seventeenth century log cabins were common in northern Europe from Norway to the Volga. Even earlier roots have been found in Stone-Age European lake villages, where some houses used crossed logs, overlapping at the corners. But it was not until 1638 that the round-log house reached the Western Hemisphere; more precisely, it reached Delaware Bay with the founding of New Sweden, roughly midway in time between Columbus and Lincoln.

Log-house construction spread, but slowly and not in all directions. The English and Dutch ignored it. New

Architect John Bullock published *The American Cottage Builder* between the campaigns of William Henry Harrison and Abraham Lincoln. He was concerned with the future building of homes in America.

England and Virginia continued to build English medieval cottages, with more elaborate English manor houses for the affluent. In Philadelphia brick walls went up over timber framing. But German and Scotch-Irish frontiersmen were carrying the log cabin west. In the wilderness, the land could be cleared as the log house was built—sturdy, cozy, defensible, fast, and without the need of precious "nayles."

The true log cabin was normally a temporary wilderness shelter, serving the frontier family until there was time to build a larger, better-appointed house out of hewn logs, mortised at the corners, with a wood floor, iron hinges, a stone chimney, and glass windows. The first cabin was likely to have a dirt floor, leather hinges, a wooden chimney (if any) with clay lining, and greased paper or skins for window coverings. Roof shingles were split oak, ash, or cedar, held in place by weight poles. The cabin could literally be built with no nails or hardware, and with no tool other than an axe—but it wasn't easy. Unless good, regular pine or spruce logs were available, huge gaps between logs had to be chinked with chips, sticks, straw, moss, clay, mud, or some combination of these. Notching the logs for a tight fit and plumb corners was a special skill that the early settlers, except for the Swedes, simply did not have.

Proliferation of log cabins began in earnest around 1700 and soon predominated in most frontier settlements. After the Revolution, a new nation was ready to romanticize its dramatic wilderness origins. Since publicity was still a fledgling art, it took a little time; but in 1840, when a Democratic newspaper sniffed that William Henry Harrison would be better suited to a log cabin than to the White

Whig propaganda aimed to present presidential candidate William Henry Harrison as a simple Westerner with log-cabin values. In this 1840 engraving, he greets a wounded veteran of Tippecanoe.

House, the Whigs seized on the slur and turned it into the famed "log cabin and hard cider" imagery of The "Tippecanoe and Tyler too" campaign.

That campaign launched the legend that the lore of Abraham Lincoln would climax twenty years later. Two centuries of architectural history came unstuck, and before long every schoolboy could tell the story of the Pilgrim wading ashore, axe in hand, and inventing log-cabin technology on the spot. Older boys—scholars, historians, politicians, illustrators, and the planners of centennials—filled books, speeches, articles, and pageants with the same myths well into the twentieth century.

During the hubbub of the Harrison campaign men like Downing and Bullock shared the country's preoccupation with humble dwellings, but not with nostalgia. Their eyes were on private visions of a future in which they was attractive homes for the poor as well as the rich. For the rich Downing had designed landscaping and gardens along the Hudson, and he had planned the grounds for the Capitol, the White House, and the Smithsonian. Now he designed—and Bullock compiled the designs of others—for everyone, with cottages that started at $200.

From this distance a few charming and perfectly practical cottage plans appear to be buried in a gallery that ranges from the curious to the downright eccentric. But in looking at a few of the designs from *The American Cottage Builder*, it is well to remember that most of these plans had already been built, in one nonplussed neighborhood or another.

Early colonists with different backgrounds favored stone, thatch, mud brick or stone construction in addition to logs. When logs were scarce, bricks could be made of ordinary clay and straw. The settler could sun-dry bricks and be ready to start construction in 10 or 12 days.

The model cottage

"I hope the day is far distant when Prince Albert's Model Cottages will be thought an appropriate residence for an American laborer. The room is too confined, the size is too small for our people." With this scolding, Bullock presents the model cottage erected by Queen Victoria's consort for the great Crystal Palace exhibition of 1851. Planned for four families, the structure enclosed a living room, bedroom, and pantry for each. Albert, always alert for ways to apply science to the improvement of society, used the models to promote hollow bricks for insulation, cast-iron springers, and wrought iron tie-rods.

Thatched cottage

"Very suitable for lodges," the thatched cottage is offered as an apt blend with the scenery of the country. "Straw thatching," Bullock points out, "is a covering easily provided in any agricultural vicinity, and is capable of being repaired from time to time at trifling cost: but it is easily accessible to vermin, and, therefore, objectionable, for the better class of cottages." Instead, he recommends reed, to be woven by experienced thatchers, with creeping plants spread over the surface of the roof. All told: a living room, kitchen, two bedrooms, veranda, tool house, and outbuildings for $400 to $900. Columns are unbarked oak; occasionally upper branches were left on two adjacent columns to form an arch.

The cabin could literally be built with no nails or hardware, and with no tool other than an axe—but it wasn't easy. Unless good, regular pine or spruce logs were available, huge gaps between logs had to be chinked with chips, sticks, straw, moss, clay, mud, or some combination of these.

Rural home no. 1

Attributed to Swiss influence, this cottage is suggested for rural or suburban settings along the Hudson, where Downing had designed and landscaped many homes. Ideally, the author says, it will be surrounded by greenery and orchards, with "the luscious peach, with its downy cheek, the buttery, melting pear, the red-cheeking cherry, the golden apricot; and, peeping out, nearly smothered by its own habiliments, the never-cloying delicacy of the strawberry, as if to escape that inevitable destiny peculiar to strawberries, of being smothered in cream." The cottage? "Perfectly simple in form—the variety in the external elevation being produced by the projecting roof." Estimated cost: $1,300.

Cottage of the Society for Improving the Condition of the Poor

For an estimated $800 to $1000, this duplex would provide each occupant a living room "fourteen-six by fifteen feet...thus affording the utmost possible space for the little furniture which the tenant may have to arrange in his 'best room.'" A scullery and pantry are attached in the back, which is actually the front since it also includes the entrances and a three-foot-square lobby for each. "The aid of the cottager would of course be expected," the architect notes, "to train honeysuckle or whatever he most desires along its frontage."

Gothic suburban cottage

Listed as the residence of C. Prescott, Esq., Troy, New York, this house was designed, Downing is led to believe, "by his intelligent lady, Mrs. Prescott, who, not being an architect, called to her aid H. Thayer, the Architect by whose united labors the designs and plans were completed." Apparently uncertain whether the word "cottage" quite suits an eighteen-room, $15,000 residence, Downing elsewhere calls it a "villa."

The village cottage

This design is described as "much recommended and adopted in England" and as "suitable for mechanics." Stone walls are eighteen inches thick. The first floor consists entirely of a 15- by 13-foot living room, a 19- by 13-foot kitchen, and a small pantry beneath the stairs to the bedrooms above. Out of sight in back is a lean-to building against the kitchen, "divided into a washhouse or scullery and fitted up with the usual conveniences, and a tool house and the requisite out-offices."

Suburban octagonal cottage

William H. Wilcox designed this octagonal house and urged its advantages over rectangular and "irregular" plans: "The octangular cottage encloses one-fifth more superficial area (with a given extent of external wall) than the square," Wilcox wrote. "It offers less resistance to the wind, looks equally well from all points of view, and furnishes an abundant supply of closets. It can be built well (if of wood) for about $1,500."

Stretching the word "cottage" a bit, Downing presented an 18-room Gothic manor house. Less expensive models included the Village Cottage for workers, an octagonal house (floor plan at right) and an Italian Cottage recommended for commanding sites.

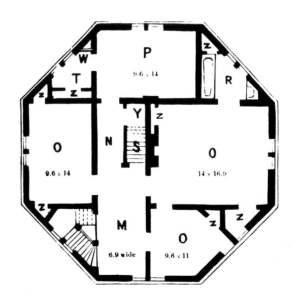

Italian cottages

Not to be wasted in valley or town, the Italian cottage is suggested for "a position visible from a considerable distance, and commanding from its site an extensive view." The tower is "eight feet and a half high," to be used as a porch on the ground floor and a bedroom upstairs. Estimated cost of the cottage: $400 to $900 in 1850 currency.

Byzantine cottage

In a neat piece of circular reasoning, Bullock dismisses the Byzantine cottage and omits further description: "The Byzantine Cottage does not ordinarily suit the general observer," he reports. "The reason is, because it is Byzantine, and is in this country out of place." He does include a cost estimate: $2,500.

Prairie cottage

The direct competitor of the log cabin at the time Downing and Bullock were writing was this frontier cottage of unburnt brick: "peculiarly adapted to settlers in the Western States." (For an unburnt brick recipe, see the next page.) Planned size is 28 by 18 feet, to include a 16-foot-square kitchen-washroom-pantry-living room and two small bedrooms on the first floor. Outer walls rise three or four feet above the joists to make a loft large enough for "a lodging room over the bedroom, or it may accommodate boarders."

How to make unburnt brick

Henry Leavitt Ellsworth (Yale, class of 1810) was a lawyer, U.S. commissioner of patents, and close friend of Washington Irving. He was also the author of a publication entitled *Plan for Cheap Cottages*, in which he instructs the frontier farmer or rancher on making bricks for a house such as the Prairie Cottage shown here:

Select a suitable spot of ground, as near the place of building as practicable, and let a circle, ten feet or more be described; let the loam be removed and the clay dug up one foot thick; or, if the clay is not found on the spot, let it be carted in to that depth. Any ordinary clay will answer. Tread this clay with cattle, and add some straw cut six or eight inches long—using two common bundles to one hundred brick. After the clay is tempered by working it, the material is duly prepared for the brick. A mould is then formed of plank, of the size of the brick desired.

In England, they are usually made eighteen inches long, one foot wide, and nine inches thick. I have found the most convenient size to be one foot long, six inches wide, and six inches thick. The mould should have a bottom not air-tight, since mortar will not fall when a vacuum is produced. The clay is then spread in the moulds in the same manner that brick moulds are ordinarily filled. A wire or piece of iron hoop will answer very well for striking off the top. One man will mould about as fast as another man can carry away—two moulds being used by him. The bricks are placed upon the level ground where they are suffered to dry for two days, turning them edgewise the second day; and then packed in a pile, protected from the rain ,and left ten or twelve days to dry.

For mortar Ellsworth recommends three parts clay, two parts ashes, and one part sand—also usable for plaster if lime is not available for lime plaster.

A Cherokee woman sits outside a log house in the Oconaluftee Indian Village in Cherokee, North Carolina. The village recreates an 18th century Cherokee community.

The Whigs, seizing on a political misstep, in 1840 presented their candidate William Henry Harrison as a simple frontier Indian fighter, living in a log cabin and drinking cider, in sharp contrast to an aristocratic, champagne-sipping Van Buren. Harrison was, in fact, a scion of the Virginia planter aristocracy. He was born at Berkeley in 1773. He studied classics and history at Hampden-Sydney College, then began the study of medicine in Richmond.

The First Houses

Neolithic notes on urban planning

In attempting to connect the past with the present, some explorers work against tighter deadlines than others. The paleontologist can rummage for specimens safely preserved in tar or fossilized in rock for thirty million years. The archaeologist knows that if his grant doesn't come through this spring, a town or tomb buried for five thousand years can probably wait one more.

The cultural anthropologist, on the other hand, has to get there while the poles are still wet, the fish are still frying, and someone is ready to tell stories after supper and exhibit some "behavior," all before the arrival of the highway department or the transistor radio. Doomed to lose many such races, scholars in the field of "urgent anthropology" rush to catalog primitive cultures before they are modernized beyond recognition. Unrecorded language, customs, traditions, and folklore can simply evaporate.

Dwellings are more durable, but as the least monumental of architecture they are the most vulnerable to extinction. The marvelous thing about primitive houses—unless you live in one—is that although they don't last long, they do persist. Most of the kinds of shelter built before and throughout history are still being built—in some fashion, in some region. And where houses are clustered in villages, towns, and cities, most of the municipal problems of Jericho or Carthage will eventually turn up in Muncie.

Facades

Since prehistoric times, food storehouses have been invested (and sometimes adorned) with personalities whose moods signify prosperity or famine. Legend has it that the stilted granaries of the ancient Iberians used to come to life and dance in the night. Some of their anthropomorphic cousins in Africa are still smiling.

Above: Beehive dwellings in North Syria make up a sizeable community.

Opposite: Conical thatch roofs over mud walls are a common and durable design in many parts of Africa.

The Mesa Verde cliff dwellings represent a massive construction project, yet the Anasazi lived in them only about 75 to 100 years—by 1300 AD they had abandoned the area.

The Cliff Palace in Mesa Verde National Park contains 217 rooms and 23 kivas and had a population of 200 to 250 people.

Master plan: transportation, refuse disposal, and food supplies

Few planners would care to lump these categories together, but consider the *de facto* master plan of pole-hut villages built over water. These began in the Late Stone Age and still exist in the marshes of Cambodia and New Guinea and the inner reaches of the Amazon. Transportation is by water. Garbage disposal is into water. And a good part of the food supply comes out of the water. Nor do residents have far to paddle for hunting and fishing: the refuse they throw into the water attracts marsh fowl and fish to the village.

Condominiums

Within a country which thinks of itself as beginning its third century are pueblo Indian tribes (including the Pueblo tribe

This Stone-Age dwelling at Skara Brae was built in 3100-2500 BC.

In a Hausa village where polygamy is practiced, each wife has a separate hut within the family's living quarters. The largest hut is for wife number one.

Uncommon Structures, Unconventional Builders

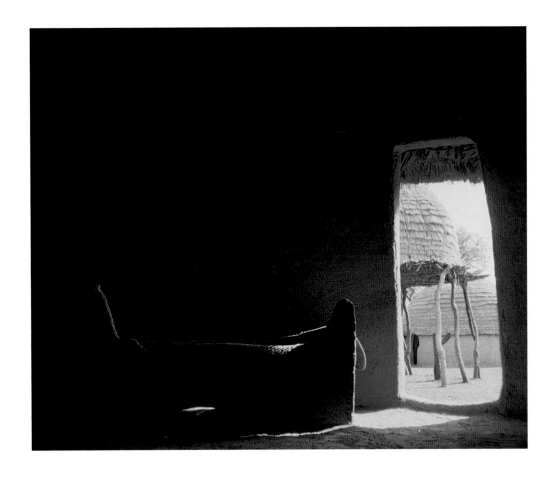

The interior of this Hausa house in Nigeria is made almost entirely of mud, including the bed next to the doorway. Just outside the door is a grain store raised off the ground to protect the crop from animals.

One culture erects golden arches over its hamburger stands; another adorns its granaries to draw good luck or friendly spirits to the harvest. These Dogon granaries in Kenya are decorated with masks—the white stains in the upper right of the photograph are from blood shed in past sacrifices.

The First Houses: Neolithic notes on urban planning

In some climates, living underground is the best way to stay cool. This underground house is in Tunisia.

Storage huts in Cameroun are raised off the ground to protect the contents from foraging animals and insects.

Opposite: A house of dried mud in the old part of Kano, on of the major Hausa-Fulani city-states of Northern Nigeria.

itself) whose people have been cultivating crops for forty centuries and building towns for fifteen. Their settlements first honeycombed the Southwestern hillsides with subterranean apartments, then gradually evolved the communal architecture of the pueblos. Examples of both survive in the tableland of Mesa Verde in southwest Colorado (now a national park), showing development of clay and stone pueblo clusters by 800 A.D.

None of this was "discovered" until a hundred years ago. The first small cliff houses were noted in a survey taken in the 1870's and in 1888 two startled cattlemen found "Cliff Palace," a multistoried complex of more than 200 rectangular apartments and circular ceremonial chambers.

In the meantime, the U.S. Bureau of Ethnology had responded to the first discoveries with an 1885 expedition sent to explore ancient dwellings of the Supai and Santa Clara Indians. After digging

In the 6th century AD, the Anasazi, or "Ancient Ones," established villages on the high, flat land in southwestern Colorado. At first, they lived in caves and in low shelters built over pits on the mesa top. In the late 1100s they began constructing multistory stone apartment houses or pueblos tucked into ledges and under rock overhangs. Some 3,800 sites have been recorded, including one particular complex, today called the Cliff Palace, which is made up of more than 200 rooms.

The First Houses: Neolithic notes on urban planning

American Plains Indians packed tools and utensils in an animal hide tepee cover and dragged it on wooden runners to their new hunting grounds.

into a series of caves carved out of volcanic rock, the group glumly reported evidence "piled two and three feet thick" that the caves were stables, not dwellings.

Multifamily dwellings

Remains have been found of Stone-Age villages with houses large enough for five to twenty families. Sometimes there were separate dwellings for women and children and large dormitories for men, an arrangement that persists in some parts of Africa, Borneo, and Melanesia. In the highly developed lake villages of the Bronze Age, rectangular pile dwellings often covered an area of 80 to 100 square yards. Space was divided into a large, open platform and various rooms, including a kitchen and workroom and a living and sleeping area with a fireplace, beds, and often a loom.

Zoning variances

Where polygamy is practiced, distinctions between single-family and multifamily dwellings tend to blur. In one version of the African *kraal* there is a separate hut for each wife and her children, enclosed in a palisade along with the master's house, to which the favorite wife has access and other wives are occasionally invited.

Population shifts

Each nomadic culture develops its own brand of portable shelter. The Bedouin tent consists of a ridgepole supported by sticks and covered with a cloth made of goat hair. The "tipi" or "teepee," used by the Dakota, Cree, Chippewa, and others, served as both shelter and luggage. For travel, the Indians wrapped their utensils, tools, and weapons in the tipi cover and often dragged it from place to place using the tent poles as a skid. Perhaps the simplest of all temporary shelters is the

The Mesa Verde cliff palaces were inhabited by Anasazi people and have survived the elements to stand as a well-preserved remnant of a civilization that flourished a thousand years ago.

Uncommon Structures, Unconventional Builders

Above: Reed granaries in Malawi exhibit conical thatched roofs, along with interlaced fiber siding.

Left: In a Mmemo Village, people cut triangular windows to let light into their mud and thatch dwellings.

Above: This village near Chiang Mai, Thailand, has palm thatched roofs with wooden supports.

Right: According to Edward Curtis, photographer and scholar of Native American culture, Assiniboine tipis were sometimes painted with dream symbols designed to bring their inhabitants good luck.

Uncommon Structures, Unconventional Builders

Right: Bedouins in the Middle East are some of the only nomadic people that remain today. The spare design of their dwellings and the small amount of personal belongings make the life more viable.

Below: This 17th century Jesuit mission in Ontario, Canada, mimicked the architecture of the native people.

"benab," an instant hut made by roving hunters in Guyana. They lay a few palm leaves out flat, bind their stalks together, and stick them in the ground. The leaves form an arched shelter roof.

Semang pygmies of the Malay Peninsula and a related tribe on the Andaman Islands in the Bay of Bengal have another version of the portable hut. They simply prop up a plaited mat and sleep to leeward. In more permanent camps, this mat is raised on four posts to form a slanted roof, or two mats are joined in a saddle roof. The Andamanese place these huts in a rough circle around a cleared dancing ground.

In a Hausa village grass thatched roofs are supported by low walls or stilted frames. The roof is sometimes constructed on the ground and lifted into place.

The Great Oval Salon and the lions in the garden at Vaux-le-Vicomte typify the extravagant details that can be found in every aspect of the chateau.

Fouquet's Folly

Certain pitfalls to avoid in building a $15 million house

Nothing was too good for Nicolas Fouquet. As the second most powerful man in France, he required a country estate suitable to his station. As a man of superlative taste and a patron of the arts, he was able to assemble for the task the most gifted architects, artists, sculptors, and craftsmen of seventeenth-century France. And as a nimble superintendent of finance under Louis XIV, he had his construction budget wired to the national treasury.

The result was Vaux-le-Vicomte, 50 kilometers southeast of Paris, a magnificent château whose planning was so perfect that history recalls only one flaw: it was better than the king's.

Louis XIV was twenty-three years old. His activities to that time had been largely concerned with horses, hunting, and girls, while the running of France had been left to the young king's preceptor and chief minister, Mazarin.

When Mazarin died in March of 1661, it was Nicolas Fouquet who inherited much of his power and had every expectation of acquiring it all. Born of the lesser nobility, Fouquet had risen to the titles of Viscount of Melun and of Vaux, Minister of State, Superintendent of Finance, and Procurer General.

He was renowned as a financial wizard, a wealthy man by virtue of two matrimonial masterpieces, and the most surefooted traveler through the labyrinth of loyalties and venalities that characterized the French Court.

Fouquet's stature can be measured by his guest list. His party in the summer of 1661 to exhibit the new château at Vaux was attended by the entire Court, including the king, who had to don layer upon layer of ruffles and lace and endure a

During his time, Nicolas Fouquet was the second most powerful man in France. Vaux-le-Vicomte was constructed as his country home and cost nearly $15 million to build.

157

An aerial view of the chateau offers perspective on the immense size of the house itself. This magnificent house inspired Louis XIV to build a new house, as well: Versailles was the result.

Fouquet was renowned as a financial wizard, a wealthy man by virtue of two matrimonial masterpieces, and the most surefooted traveler through the labyrinth of loyalties and venalities that characterized the French Court.

three-hour carriage ride from Fontainebleau under an August sun.

Louis was soon to forget the heat. What he encountered at Vaux-le-Vicomte was enough to recall more substantial annoyances—enough to make a man grow suddenly tired of living at the Louvre and summering at Fontainebleau, and enough to make an ambitious young king wonder if the royal treasury was any longer big enough for two.

For the building of Vaux, Louis and the taxpayers of France had unwittingly furnished at least a million *livres*, at least fifteen million dollars in today's currency. Not that Fouquet was doing anything unusual; other finance ministers had siphoned before him. But none of his predecessors had made such a spectacular display of their profits.

Uncommon Structures, Unconventional Builders

The King's entourage approached the château through a vast forecourt laced with broad avenues, manicured groves and hedges, out-buildings faced with matching stone, finally a moat, and just beyond it, the house itself: a massive structure designed for Fouquet by Louis Le Vau, who was later to serve as architect for the palace at Versailles.

Huge, ornate, bristling with mansards (Mansard himself was still alive), embellished with statuary, and crowned by a sixty-foot dome, Vaux mated French château traditions with a strong Italian influence. Inside, the design genius of Le Vau blended with the artistic genius of Charles Le Brun, who had drained the reservoirs of French talent to assemble a task force of nine hundred artists, sculptors, and craftsmen for the interior-decoration project. In nearby Maincy he had built an entire furniture and textile factory, specializing in tapestries, just to supply the walls and halls of Vaux-le-Vicomte.

Louis toured the grand salon under the dome, a cluster of huge tributary rooms around it, and a series of apartments—any one of which would have put the king's own quarters to shame. Friezes, tapestries, endless sculpture and relief, decorative fixtures and molding were just the beginning. The walls and ceilings were alive with epic after epic in mural and fresco—several acres of allegory and symbolism—climaxed by Hercules on Olympus. It wasn't entirely clear whether Hercules was meant to symbolize the king or Fouquet.

Like too much candy, the beauty might have become unbearable; but Louis got no relief when he stepped out onto a balustraded walkway overlooking the gardens. More magnificence. André Le Nôtre,

---

Even the back door is spectacular—from the rear entrance, Vaux-le-Vicomte is just as imposing.

The chateau of Vaux-le-Vicomte owes its name to the convergence of two valleys on a noble estate.

Fouquet's Folly: Certain pitfalls to avoid in building a $15 million house

André le Nôtre, designer of the Tuileries, devised an ingenious system of perspective for the east gardens of Vaux. Transverse axes of water connecting the garden's 250 fountains create an optical illusion that makes the chateau seem nearer.

Left: The death of Cardinal Jules Mazarin in 1661 opened the door for Fouquet to take over vast amounts of wealth and power.

Opposite: No one was more impressed by Vaux than Louis XIV—upon seeing the chateau, he hired the designers to build Versailles.

illusion that—although the foreground kept receding as he moved away from the house—the house itself seemed to remain near. But from where the king stood, the scale alone was overwhelming. Beyond moats, groves, grottoes, terraces, and cascades, he could see a towering statue of Hercules, the crown jewel of the gardens. Hercules was two miles away.

If the architecture and ornamentation of Vaux were overdone (the king thought so), housewarming festivities supplied the *coup de grâce*. Their cultural equivalent today would have to include a serenade by Leonard Bernstein, commemorative verse by Robert Lowell, theatricals by Stanley Kubrick and Neil Simon, catering by Lutece, and a massed salute by the Seventh Fleet.

For Fouquet it was dinner by Vatel, the finest chef in France, served to music composed and conducted by the country's most prominent musician, Lully. Then a new play written by (and starring) Molière, poetry by La Fontaine and Pellisson; and finally an apocalyptic display of fireworks by Torelli of Italy, climaxed by exploding *fleurs-de-lis* and Roman candles from the rooftop, to the accompaniment of trumpet and cannon.

Three weeks later Fouquet was arrested by the musketeer D'Artagnan (later celebrated by Alexandre Dumas, but for other feats). The charge was embezzlement, and the warrant was signed by the king. If Fouquet's detractors in the Royal Court had not been able to convince Louis that the treasury was bleeding, Vaux *had* convinced him. Then too, the young monarch had something else on his mind: Versailles.

Before poor Fouquet was even convicted, Louis was having shrubbery and statuary hauled from Vaux to the site of his own great construction project.

Peopled with statuary in marble and gilded lead, the gardens boasted at least 50 major fountains and scores of sparkling jets and sprays.

designer of the Tuileries, had now designed the gardens of Vaux. Peopled with statuary in marble and gilded lead, the gardens boasted at least fifty major fountains and scores of sparkling jets and sprays.

Fountains and waterways were placed by Le Nôtre according to a private geometry that defied ordinary perspective. For example, he used transverse axes of water to reflect sunlight in such a way that a visitor walking out into the gardens had the

Nicolas Fouquet was the superintendent of finance under Louis XIV and was accused of diverting public funds and plotting against the king. He died in the prison of Pignerol in 1680.

Fouquet's Folly: Certain pitfalls to avoid in building a $15 million house

Dazzled by Vaux, Louis had decided to dazzle the rest of Europe. Before he was through, nearly every feature of Vaux had been incorporated into the new palace at Versailles, only bigger.

Dazzled by Vaux, Louis had decided to dazzle the rest of Europe. Before he was through, nearly every feature of Vaux had been incorporated in the new palace at Versailles, only bigger.

Louis expropriated not only the ornaments of Vaux but its geniuses as well. Le Vau was his architect; Le Brun designed his interiors; and Le Nôtre laid out his gardens. Virtually the entire community of artisans working at Maincy moved to Versailles. And when the new palace was ready for its first feasting, Molière was there with a play, Vatel with a dinner, La Fontaine with a poem, Lully with a recital, and—sure enough—Torelli with his fireworks.

As for Fouquet, he had too many admirers to be executed as Louis had intended. He was imprisoned for sixteen

On August 17, 1661, Fouquet entertained Louis XIV at Vaux-le-Vicomte with so lavish a party that the king is said to have threatened to have him arrested.

Above: The palace and gardens of Versailles dwarfed Vaux-le-Vicomte—but their design was derivative of the innovations and details of the smaller chateau.

Opposite: Charles Le Brun, shown here in a portrait by Nicholas de Largilliere, assembled a team of almost 1000 painters, artists, and craftsmen to complete the interior of Vaux-le-Vicomte.

years, dying just before he was to be set free. And Vaux-le-Vicomte, after many generations and two restorations, stands today very much as it was in 1661.

Its preservation is fortunate: there will never be another like it. Today's architect cannot monopolize a nation's artists and craftsmen, as Le Vau and Le Brun could. Nor can he build an on-site tapestry factory, nor arrange the topless budget that Fouquet enjoyed with his direct, unmetered pipeline from the treasury. For that matter, where would an architect in this century find a client willing to foot the water bill for Le Nôtre's 250 fountains?

Fouquet's Folly: Certain pitfalls to avoid in building a $15 million house   165

The Globe Theatre was restored to working order in a massive 1994 reconstruction.

Confusion Around the Globe

The cloud-capp'd towers, the
*   gorgeous palaces,*
The solemn temples, the
*   great globe itself,*
Yea, all which it inherit, shall
*   dissolve*
And, like this insubstantial
*   pageant faded,*
Leave not a rack behind.
*   We are such stuff*
As dreams are made on...

The Tempest, Act IV, Scene I

In April of 1597 the lease ran out on a small piece of land in the Shoreditch section of London. The landlord, one Giles Allen, refused to renew on the grounds that his tenants' activities had been boisterous, unruly, and vulgar. Some historians think Allen had a more compelling reason: with the tenants evicted, he stood to acquire their building—a large wooden theater called The Theatre.

The name was not as unimaginative as it may sound. When it opened it 1576, The Theatre had indeed been *the* theater in London and perhaps the first public playhouse in English history. The builder—Allen's tenant—was James Burbage, an actor in the days when players had to perform as best they could in the inn yards along Gracious (now Grace-church) street.

Once Burbage had set the example, the great wooden "plaie howses" of

William Shakespeare, in a painting by poet and artist William Blake.

According to John Russell Brown in *Shakespeare and His Theatre*, "one rough sketch with a few characters standing in a row is all the pictorial evidence we have suggesting what actors looked like during an original performance of one of Shakespeare's plays."

167

The Globe

Uncommon Structures, Unconventional Builders

Opposite: Historians have had to sort through numerous illustrations to piece together what the original Globe looked like—this drawing, though not to scale, is accurate in its depiction of the theater's octagonal shape.

Below: The 1994 reconstruction made the Globe a popular tourist attraction—today, visitors tour the theater and even watch plays from its benched galleries.

London proliferated: the Swan, the Rose, the Fortune, the Hope, and the Curtain. Each housed an accomplished repertory company with its own playwrights, plus at least one quick-penned copyist to pirate plays from the other theaters. The actors would perform in their playhouse during fall, winter, and spring, then go on summer tour in the country when hot weather and city crowds aggravated the plague.

With his competitors flourishing and his lease about to expire, Burbage built a new and even finer theater on the west side of London, converting a building once used by the monk of Blackfriars into England's first roofed-in, lighted public theater. But Blackfriars was in an exclusive residential enclave whose occupants did not care for the idea of drums, cannon,

Also according to Brown, "the word 'globe' had entered the English language less than fifty years earlier to refer to spherical models of the world that was newly being discovered."

On Bankside, in a section called the Liberty of Clink in Southwark, the syndicate used its salvaged timbers to erect the finest playhouse in London and the premiere house for most of Shakespeare's plays: the Globe.

trumpets, and the "vagrant and lewd persons" that the performance of plays might attract. They petitioned the Privy Council and tried to have Blackfriars suppressed.

Burbage died less than two months later. His sons, Richard and Cuthbert, reacted differently. They contacted several friends, including a builder named Peter Street and a young actor and playwright named William Shakespeare. With Blackfriars in trouble and The Theatre in the hands of the landlord, the Burbages formed a syndicate and attempted to negotiate an extension of the lease. When that failed, they took matters into their own hands.

Three days after Christmas of 1598 the Burbages, Shakespeare, four other actors, and some hired help arrived in Shoreditch with wrecking bars and, under Street's expert direction, commenced to dismantle The Theatre and cart off its timbers across the Thames to Bankside.

Historians, architects, and artists collaborated to interpret the sketches and drawings of the Globe's interior in order to produce the stage design that stands today.

Prior to the construction of permanent theaters such as the Globe, acting companies traveled from place to place performing plays in town buildings, private homes, or from portable platforms.

Uncommon Structures, Unconventional Builders

The Shakespeare Memorial Monument in London is a tribute to Britain's most famous literary figure. Shakespeare's own words, carved on the side panels, pay homage to the Bard: "Good night, sweet prince—and flights of angels sing thee to thy rest."

CONSIDERATION·LIKE

AN ANGEL CAME,

AND WHIPT THE OFFENDING

ADAM OUT OF HIM.

GOOD NIGHT, SWEET PRINCE

AND FLIGHTS OF ANGELS

SING THEE TO THY REST

THIS MONUMENT

During Elizabethan times, it was unacceptable for women to act on stage, so boys and young men played the female roles.

This giant pillar at the base of the Globe's stage has been called the 'pillar of Hercules.'

(Allen later claimed that, in addition to wrecking bars, the company also brandished swords, daggers, and axes and behaved in a "very riotous, outrageous, and forcible manner." He sued for damages and lost.)

On Bankside, in a section called the Liberty of Clink in Southwark, the syndicate used its salvaged timbers to erect the finest playhouse in London and the premiere house for most of Shakespeare's plays: the Globe.

Few structures have ever received the scholarly attention that has been lavished on the Globe. Careers have been poured into the minutiae of circumstantial evidence that hints at its size, shape, construction, interior, stage design, and the relationship of these to Elizabethan theatrical technique. There are many sources of indirect evidence, including inferences drawn from the stage directions in Shakespeare's plays. But the prize exhibits are missing: no detailed drawings of the exterior have been found; cartographer's sketches of the period are disparate and short on particulars; there are no known drawings made from inside the theater; the builder's plans have been lost; and the Globe itself has been lost twice.

The first destruction of the Globe was by fire in 1613. One contemporary account puts it this way:

Upon St. Peter's Day last, the playhouse or theatre called the Globe, upon Bankside, near London, by negligent discharge of a peal of ordnance close to the south side thereof, the thatch took fire, and the wind suddenly dispersed the flames round about, and in a very short space the whole building was quite consumed; and no man hurt: the house being filled with people to behold the play, viz. of Henry the Eight.

By this time, the Globe's owners were the wealthiest actors in London, so they had no trouble financing its reconstruction in even grander style. The second Globe lasted three decades, then withered under the frown of a hostile Puritan Parliament. All playhouses were closed by law in 1642—twenty-six years after Shakespeare's death—and the Globe was soon torn down.

Uncommon Structures, Unconventional Builders

Following generations showed recurrent interest in Shakespeare's plays and poems but little in his theater. In 1833, near the end of the Romantic Period in literature, the *Penny Magazine* of London paused briefly to recall the Globe in a badly muddled reminiscence and to mention that its site was then occupied by a brewery. From that time through the Victorian period, the *London Quarterly Review* published dozens of essays on Shakespeare's works but not one speculation about the Globe. Modern scholarship has repaired part of this lapse; but to this day no one has precise knowledge of the Globe's design, and many historians believe that no one will *ever* know—that no decisive piece of evidence is likely to turn up.

If it did—if, miraculously, Peter Street's plans were unearthed—they might furnish structural specifics like these, taken from surviving plans for the Hope theater, drawn up in the same year that the Globe burned:

Make the principalls and fore front of the saide plaie house of good and sufficient oken tymber, and no furr tymber to be putt or used in the lower most, or midell stories, excepte the upright postes on the backparte of the said stories. (All the byndinge joystes to be of oken tymber.)

And they might supply interior detail, as does this excerpt from an agreement for construction of the Fortune:

All the princypall and maine postes of the saide fframe and stadge forwarde shalbe square and wroughte palasterwise with carved proporcions called Satiers to be placed & sett on the topp of every of the same postes.

It's generally assumed that the Hope was smaller than the Globe and differed considerably in design. The Fortune was large but rectangular; the Globe was octagonal. Or was it cylindrical? No doubt its basic shape was the same as that of its antecedent, The Theatre (since the same timbers were used); and the design of The Theatre was copied from the bear-baiting arenas south of the Thames. Or was it an elaboration of the inn yards on Gracious street? Or a fur-

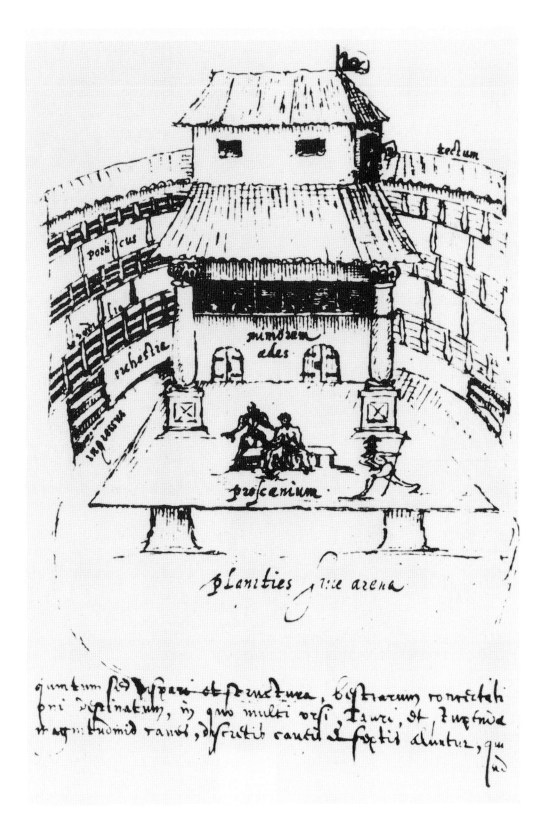

The Swan Theatre, across the street from the Globe in Southwark, had much the same design as the original Globe. Both were based on Greek and Roman designs, but not copied—one likely influence was the innyards where traveling troupes performed before the advent of theaters in England.

Confusion Around the Globe

The overhead structure was referred to as The Heavens, and its underside apparently was painted with suitable celestial symbols.

Queen Elizabeth apparently visited the Globe in 1560, to watch a production of *The Merry Wives of Windsor.*

ther development of the outdoor stage at Ghent? Or an original architectural species? Or a child of Renaissance theaters in Italy and France? Or could it be that James Burbage had direct knowledge of the Latin treatise by Vitruvius on Classical theater design? All of these possibilities have been seriously advanced and seriously challenged.

It seems fairly well established that both The Theatre and the Globe had a central yard, unroofed and surrounded by three tiers of roofed galleries, a stage that extended into the yard, a structure above the stage for machinery, and "tiring house" quarters behind the stage for dressing rooms and storage of costumes and props.

The overhead structure was referred to as The Heavens, and its underside apparently was painted with suitable celestial symbols. Machinery included a hoist to lower gods, goddesses, ghosts, or whatever figures might properly descend from heavens. For the lower end of the

supernatural spectrum, there were trapdoors in the stage, out of which devils could emerge to appropriate hissing and occasional fireworks. There was little use of painted scenery but heavy use of decoration. By some accounts the wooden columns in the Globe were painted to resemble marble and the stage area was brightly painted and festooned with garlands, tapestries, and banners. The audiences loved bright colors, rich costumed, spectacular scenes, and special effects.

There is considerable evidence that the elaborate design and paraphernalia of the stage was by no means equalled in the accommodations for playgoers. Although seating estimates vary widely, it seems likely that most of the audience at any well-attended play was crammed in on hard gallery steps with no backrests, much less armrests, while the groundlings in the central court were "closely pestered together" under the afternoon sun. (Estimates of space per person vary from

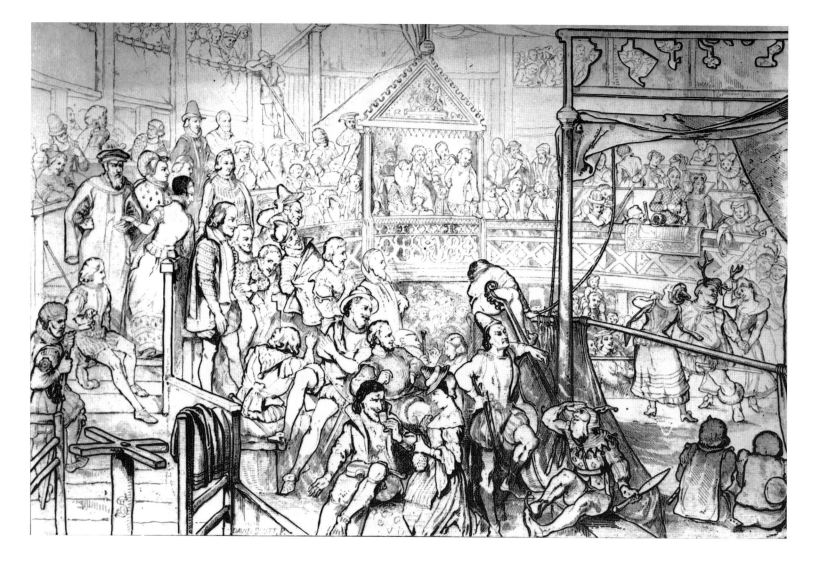

Uncommon Structures, Unconventional Builders

The Globe (right) and the Swan, in an 1825 English painting.

fourteen to twenty-two inches in seating width and from twenty to thirty inches in depth, with something between one-and-one-half and three square feet per person for standing room in the yard.) More expensive seating flanking the stage may have been more commodious, with rush matting on some of the benches. At any rate, these seats were in the shade, while many spectators in the penny galleries opposite the stage had the sun in their eyes.

Probably the best and most thoroughly detailed reconstruction of the Globe was researched and built in miniature at Hofstra University by John Cranford Adams, author of *The Globe Playhouse: Its Design and Construction*, assisted by Irwin Smith. This model is now exhibited at the Folger Shakespeare Library in Washington, D.C.

Down the hall in the Folger is an actual theater, designed to the general plan of Elizabethan stages if not specifically to that of the Globe. And in various cities across the United States, in Canada, and in England, there are "Elizabethan" playhouses that, to one degree or another, evoke the balconied thrust stages, encircling seats, and some of the production features of the Globe, the Swan, and the Hope, not to mention some of the plays.

Puritan laws weren't the first things to shut down London theaters. An outbreak of plague forced theaters to close their doors during the 1952-1953 season.

Ruins are all that remain of the oldest city on earth. This cross-section of part of the excavation of Jericho shows the stout city wall and the Round Tower.

Jericho

Why did the walls come tumbling down?

Two million years before anyone started to count, something shifted in the Earth's crust and a long, narrow slice of the surface dropped a half mile straight down, leaving a scar which now contains the Jordan River valley, the Dead sea, part of the Red sea, the Gulf of Aqaba, and the Great Rift valley through East Africa.

As seen through the lens of twentieth-century hindsight, this massive fracture has been seething ever since with upheavals of one kind or another: geological, glacial, climatic, cultural, and military.

Ice age followed ice age, and while northern latitudes were locked in glaciation, stretches of the Middle East that are now all dust and desert became lands of perpetual rain—the sort of place where a man might be tempted to build an ark. When the last ice age—or if not the last, at least the latest—receded some 10,000 to 12,000 years ago, it left a fertile region, including an abundant freshwater spring a few miles northwest of the Dead sea.

Around this spring a stone-age settlement appeared, and around the settlement—a wall.

What the original inhabitants called their town, we don't know (probably not Prepottery Neolithic A, which is what archaeologists call it). When Joshua arrived with his army around 1300 B.C. , it was named Jericho, and it was already more than twice as ancient as Rome is now. At 825 feet below sea level it is one of the lowest towns in the world, and so far as modern archaeology can determine, it could also be the oldest.

Theories of collapse

What Joshua did to Jericho is well known. Exactly how he managed to do it may never be known, but there are clues enough to tempt the imagination.

The Jordan River winding through the countryside of upper Galilee.

The Bible record shows that Jericho was destroyed between 1426 and 1385 BC. Sir John Garstang says, from the uncovered evidence, that "no distortion of the case can disguise the fact that the city of Jericho fell during the latter part of the reign of Pharaoh Airerbotep III between 1400 and 1385 BC."

"So the people shouted when the priests blew with the trumpets: and it came to pass...that the wall fell down flat, so that the people went up into the city... and they utterly destroyed all that was in the city... and they burnt the city with fire..."

—Joshua, Chapter 6

Other scenarios picture a demoralized townspeople huddled within a shaky, ready-to-tumble wall which was so unstable that the vibrations of ram's-horn trumpets and massed shouting were enough to unseat it. Or they suggest that a wall designed merely to stymie foot soldiers would not necessarily be sturdy enough to support the entire population of the city if they happened to crowd onto one section of its rim to glimpse the eerie processional of warriors and priests assembling outside to call awesome powers down upon their heads.

Inevitably there have also been more bizarre suggestions—of subsonic vibrations, of laser beams supplied by extraplanetary visitors—on which the scriptures offer no comment. But, as Will Durant has observed, history is so indifferently rich that it will furnish ample evidence to support almost any theory whatever.

Why Jericho?
While the walls stood, they were formidable. The odds against Joshua must have seemed long. Remnants show that Jericho's fortifications during this period included double walls, one of them twelve to fourteen feet thick. Houses in the town leaned against it, and at least one family lived inside the wall itself.

The infamous battle of Jericho has been a popular subject in religious art for centuries. This painting, by James Jacques Joseph Tissot, is housed at the Jewish Museum in New York City.

The first major excavation of the site of Jericho, located in the southern Jordan valley in Israel, was conducted between 1907 and 1909. The puzzled excavators found piles of mud bricks at the base of the mound the city was built on. Decades later, excavators realized these piles of bricks were from the city wall, which had collapsed when the city was destroyed.

Most familiar, of course, is the "Trumpet Theory." Less familiar—since artists have ignored it and it has never been set to music—there is a "Strumpet Theory." Both have ample support in the Old Testament, which also leaves room for the most common physical explanations of the wall's collapse: an earthquake, or the simple persistence of ditchdiggers.

Uncommon Structures, Unconventional Builders

But however impregnable those defenses appeared, Joshua must have known something that gave him confidence. He knew about the walls. He knew about the town. He had sent spies who spent several days scouting it thoroughly, inside and out.

There were other cities that could have been attacked first, and a first success was imperative in the Israelites' conquest of the Promised Land. Joshua was far too astute a general to risk a costly, time-consuming failure; and yet, it was Jericho that he singled out for a siege.

From above, it is easy to see how extensive the excavations of Jericho must be—here, only a tiny portion has been cleared, as seen from the southwest.

What was Joshua's secret?

The first part of his secret was no secret at all but an obvious quandary. Joshua knew that somewhere, somehow, his army would have to breach its first wall. The land of Canaan was flowing not only with milk and honey but also with the migrations of nomadic tribes and with commerce between Mesopotamia and the Mediterranean ports. Its settlements had prospered and become inviting targets for raids by neighboring city-states and by nomads from the south and east. So virtually every town of any consequence had a wall—or two walls, or three.

A world of walls

At every strategic point the Promised Land and all the lands around it were con-

Given that Jericho is located in roughly central Palestine, access to neighboring city-states was a major key to Jericho's importance to invaders and to traders. Jericho's location was ideal for the establishing of trade routes and for communication exchange.

To the northwest of Jericho, the remains
of the Monastery of Temptation loom on

An ancient stone star symbol rests near the Monastery of Solitude, in Jericho. The star marks the place where the Israelites crossed the Jordan River to the Promised Land.

Jericho benefited from natural irrigation afforded by the Jordan River approximately four miles to the west, and from underground tributaries from the Central Mountains, which fed its famous oasis. This irrigation resulted in teeming plant life and helped to transform Jericho into a flowing sea of green in an otherwise barren desert.

trolled by key cities, and the cities were secured by walls.

Though the region is sparsely supplied with quarry stone, the Sumerians had invented the brick mold about 3000 B.C., and wall building soon proliferated. Mud bricks could be made quickly and replaced quickly. Kiln-fired bricks were a luxury because combustibles were also scarce, but the bricks could be baked hard in the sun in a few weeks or, even better, a few years.

These replaced the unstable hand-packed planoconvex or "hogback" bricks of earlier structures at Jericho and elsewhere. By Joshua's time three successive civilizations had built and rebuilt at least seventeen walls around Jericho, each felled in turn by earth tremors, invaders, or gradual capitulation to sun, wind, and rain.

To the immediate west, the city of Ai was girded by a *triple* wall (which Joshua later penetrated by a ruse—pretending to retreat in panic but leaving a hidden contingent to enter the city after the Canaanite troops came charging out in pursuit). To the north Hazor was encircled by a 24-foot-thick brick wall on stone

foundations. Gibeon, Megiddo, and Hebron had walls. Jerusalem, though most of its now-familiar structures had yet to be built, was so strongly fortified that Joshua simply skipped it in the conquest of Canaan. It remained defiantly Canaanite until the time of David, some 300 years later.

Had the Israelites ranged beyond the land of Canaan, their odds would have grown even worse. To the east lay the city of Ur, birthplace of Abraham, at the junction of the Tigris and Euphrates, where an enormous elliptical wall of baked brick stood on a rampart of unbaked brick seventy-seven feet thick. Around Ur's sister city, Uruk, a six-mile wall was studded with 900 defense towers (Jericho had one).

On the east bank of the Tigris, Nineveh would eventually surpass even this, with King Sennacherib's seven-mile perimeter of quadruple walls, fifteen gates, 1,500 watchtowers, and a 130-foot-wide moat surrounding it all. The Greek historian Diodorus claimed (though other historians doubt) that t hese walls were 100 feet high and the towers 214 feet high.

Uncommon Structures, Unconventional Builders

Nebuchadnezzar's Babylon flourished behind five concentric rings of fortifications, and the inner two were filled in between with rubble to form a combined wall 88 feet thick.

Nebuchadnezzar's Babylon flourished behind *five* concentric rings of fortifications, and the inner two were filled in between with rubble to form a combined wall eighty-eight feet thick. By the time the Babylonians added moats and crenelated parapets to this system, there was no practical way to gain entry except to go to the Ishtar gate and bribe someone.

Jericho's defenses had achieved no such splendor. Two or three foundation courses of stone were topped by rough courses of mud brick, backed by an interior facing of packed clay. There was an outer wall, six feet thick, and an inner wall roughly twice as massive. The twelve to

The original walls of Jericho are the result of seismographic activity producing a great rift extending from the Sea of Galilee to Northern Africa. The land in this region dropped some 3,000 feet and settled at least 900 feet below sea level.

Salt formations dot the coast of the Dead Sea.

Uncommon Structures, Unconventional Builders

eighteen feet between the two was open except at the northwest corner, where it was filled in to support a forty-foot watchtower.

Inside the walls

But there was something else between those walls. An accomplice. Her name was Rahab, her profession as old as the town itself; and she and her family were destined to be the only survivors of Jericho after the fall.

When Joshua had sent two of his men to scout the defenses of Jericho, it was Rahab they visited.

"All who live in this territory have been seized with terror at your approach," she told them, "for we have heard how Yahweh dried up the Sea of Reeds before you when you came out of Egypt...No courage is left in any of us to stand up to you."

These were the Canaanites, part of the Phoenician civilization, and the people who would later build Carthage and very nearly conquer Rome. But it was the Canaanites a thousand years before Hannibal, and there was no such able general to oppose Joshua or to rally the townspeople.

When word spread through the security-conscious town that there were two outsiders in Rahab's house, she hid them under a pile of flax stalks on her roof and claimed they had already left. Later she helped them to escape in exchange for the

The archaeologists' field report dated March 2, 1930 states: "The main defenses of Jericho... followed the upper brink of the city mound and comprised two parallel walls, the outer, six feet thick and the inner, twelve feet thick. Investigations show...continuous signs of destruction and conflagration...traces of intense fire are plain to see...houses along the wall were found burnt to the ground..."

promise that her family would be spared when Jericho was overrun.

Three days later the spies returned to the Israelite encampment at nearby Gilgal. On hearing their story, Joshua immediately struck camp and marched on Jericho.

Outside the walls

The odds were improving. A frightened populace, roughly 2,000 people, were apparently huddled in their eight-acre enclosure, convinced they were doomed.

Still, there were the walls.

After six days of siege nothing appeared to have changed, and the fortifications stood firm. It's possible that an earthquake on the seventh day broke the stalemate: a theory often advanced, and a credible one in view of the region's history of seismic surprises. Its timing, however, would indeed be miraculous if the walls were to collapse on cue. More plausibly, the Israelites might have arrived just after a fortuitous earthquake had left Jericho defenseless, in which case the Book of Joshua would have to be seen as a narrative stretched to glorify one of the founding fathers.

But what keeps surprising secular scholars of history and archaeology is the remarkable extent to which the Old Testament holds up, verified by other evidence time after time, as a reliable reflection of history. If the scriptures say Joshua's army circled the town and his priests blew their rams' horns and the wall fell, then that may be precisely what happened. And Rahab may be the *reason* it happened.

The art of sapping was not unknown, and Joshua would have come prepared to have his own ditchdiggers undermine the fortifications if there were any means of getting them close enough to do it.

One means would be to have some of them inside, say, in Rahab's house. Another would be to distract the defenders so effectively that the sappers could work unnoticed from the outside, or at least enter unnoticed and then go to work digging between the inner and the outer walls to undermine both.

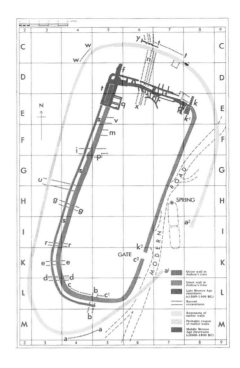

Above: The defenses Joshua had to breach at Jericho consisted of an outer wall 6 feet thick and an inner wall 12 to 14 feet thick. The city had one watchtower at the northwest corner and a gate along the East Side.

Opposite: The Mount of Temptation can be seen in the distance from the excavations at Jericho.

"All who live in this territory have been seized with terror at your approach," she told them, "for we have heard how Yahweh dried up the Sea of Reeds before you when you came out of Egypt...No courage is left in any of us to stand up to you."

Jericho had relatively easy access to southern neighbors such as Jerusalem and Bethlehem, and to northern neighbors such as Beth Shan and Nazareth.

An opening for surreptitious entry is suggested by one quirk of construction in several generations of the Jericho walls. They were sometimes rebuilt in sections, apparently to localize damage when one portion collapsed. It is not known whether Jericho's seventeenth wall had this feature but, if so, a compelling diversion on one side of the city might be more than enough to permit entry through crevices on the opposite side. And there is mention in the Book of Joshua that Rahab had agreed to hang a scarlet cord outside the wall to mark her residence.

As for diverting the attention of the defenders, Joshua had worked out one of the most elaborate distractions in the history of psychological warfare.

Six times, once a day, his army marched in solemn and silent procession around the city, accompanied by seven priests carrying trumpets and others bearing the Ark of the Covenant. One can imagine the cumulative terror of the people inside as they beheld what Isaac Asimov has called "the somber and majestic spinning of a supernatural net about the city."

On the seventh day, the procession began early and did not cease until it had circled the city *seven times*. Then it stopped. Silence, followed by the sudden blare of trumpets and a mighty shout from the whole Israelite army.

When they heard the sound of the trumpet, the people raised a mighty war cry, and the wall collapsed then and there.

The Israelites stormed the town, destroyed it, and killed everyone in it except Rahab and her family. Joshua decreed that Jericho should never be rebuilt, and he backed up the decree with a curse:

Elaborate columns and carvings are all that remain of the Hisham Palace at Jericho.

Uncommon Structures, Unconventional Builders

Cursed be the man before the Lord, that riseth up and buildeth this city, Jericho.

Curses last longer than walls.

The Israelites kept their distance from the blighted site. When they finally rebuilt Jericho, 400 years later, it was nearly two miles from the first location. This in turn was leveled during the Roman occupation, about the time of Christ. Herod sacked the city, then built it up again to include a palace. Persian invaders destroyed that. The Crusaders built yet another Jericho, this time a mile downstream from the water supply, and theirs was obliterated by Saladin. The present-day town is next to the original, still using the same freshwater spring; but the Jericho flattened by Joshua is still flat.

Nothing remains but a broad mound of rubble, a compost heap of human events where archaeologists carve careful trenches, meticulously sorting out what happened, and in what order, after the first known instance in all of history when it occurred to someone to build a wall.

The Ur Ziggurat, made of bricks and sandstone, rises out of the arid desert.

"Cursed be the man before the Lord, that riseth up and buildeth this city, Jericho. Curses last longer than walls."

The Fallen Idol

Monumental sculptures of Rhodes

n his search for building materials, man has used snow, grass, basalt, mud, paper, and pelts. But never has there been a procurement effort quite like the mustering of construction materials that took place around 300 BC in the city of Rhodes.

To defend their island against imminent siege, the citizens of Rhodes tore down some of their best architecture and converted the rubble into massive battlements. Two years later, when the battle was won, they reversed the process: they dismantled the attacker's catapults and siege machines and used the metal, plus some of the stone from their own ramparts, to erect their celebrated war memorial, the Colossus of Rhodes. The cycle was to renew itself nine centuries later, when the Saracens sold what was left of the Colossus for scrap metal—to be used for weapons.

A colossal threat

Rhodes, in the era of its Colossus, was one of the world's most beautiful, most civilized cities. Politically it had outgrown the leadership of Athens and had become an ally of Ptolemy's Egypt, at the expense of a former partner, Macedonia. Demetrius Poliorcetes of Macedonia decided to retaliate, and mounted a punitive expedition whose army of 40,000 outnumbered the entire population of Rhodes.

When the Rhodians heard of the invasion plan, they began to fortify the island. Very few stones were left unturned.

They tore down temples—many temples, because the early Rhodians had recognized many deities. They added the stone and marble to their rising battlements and melted the metal to forge weapons. And they sacrificed much of the city's magnificent public works for defense: ripping out walls for their stone,

Above: The harbor at Rhodes is a busy port, but bears the scars of history—although the Colossus has fallen, the Crusader's castle remains.

Opposite: A Roman copy of the Colossus is the closest thing art historians have to the ancient statue.

The Colossus of Rhodes was a great bronze statue, erected in about 280 BC by the citizens of Rhodes, capital of the Greek island of the same name. It represented their sun god Helios and was said to be 105 feet high.

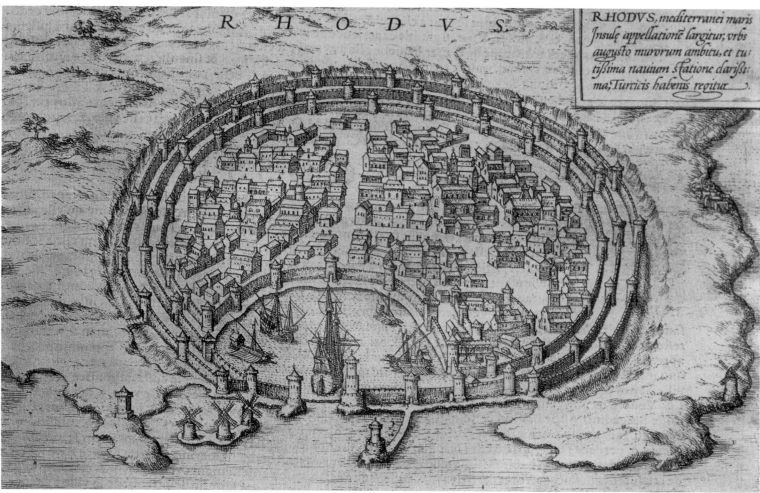

Uncommon Structures, Unconventional Builders

Opposite, top: History's second-largest colossus stood for 56 years, and its legends have grown for 2,200 years since its collapse. It stood beside the harbor of Rhodes, not astride its entrance as often depicted.

Opposite, bottom: Rhodes, in this 1573 French engraving, is shown as a heavily fortified city—three walls surround it. It's no wonder that Rhodes, the easternmost island in the Aegean, was well protected; the city was a commercial center, a naval power with colonies as far away as Spain, and a cultural center to rival Athens, Alexandria, or Rome.

sealing off roads with fortifications, and closing down one of the busiest harbors of the ancient world.

Rhodes executed one of the most thorough mobilization plans in history. Every man, woman, and child trained for the battle. Even cripples had their wartime assignments, and slaves were promised their freedom if they would fight along-side their masters.

The siege lasted over a year. Though plague-ridden and close to starvation, the Rhodians held out until a rescue fleet from Ptolemy forced the Macedonians into hurried retreat—so hurried that they left most of their formidable arsenal of war machines behind.

A monumental triumph

Leaders of Rhodes decided that Macedonian siege weapons would make an excellent monument to Macedonian defeat. They commissioned the great sculptor-architect Chares to design a

According to legend, the statue straddled the harbor entrance, but it is more likely that it stood to one side. An earthquake toppled the statue in 224 BC. Legend also says the statue broke at its weakest point—the knees.

The second largest colossus ever built still stands: the Statue of Liberty. It is a familiar sight in New York Harbor.

colossal monument to the sun god Helios (the Rhodian counterpart of Apollo) while some of his fellow citizens set to work extracting his building materials from the remnants of war.

Bronze, by this time, had come into its own. Techniques of hollow casting had arrived from Egypt or Assyria, and the

The Lindos acropolis at Rhodes is now the island's most famous ruin—built in 200 BC, it was once the Temple of Athena.

The Colossus towered at a total height of 150 feet.

metal was widely used in the sculpture of Greece and Samos. In monumental architecture, bronze was prized not only for its beauty but for its permanence, its natural corrosion resistance. By the time Chares was ready with his design, his salvage crews were ready with 500 talents of bronze (nearly 30,000 pounds) and 300 talents of iron.

Chares' plan was a daring one. The pedestal alone would stand higher than most statues known at the time—45 feet, all of white marble.

The Colossus would tower another 105 feet skyward, to a total height of 150 feet.

The Colossus was built in sections, each partially assembled on the ground and then raised into place with the aid of wooden scaffolding. It had a skin of bronze, hammered so thin that in finished form the outer shell was no thicker than a penny.

For structural strength Chares designed an iron skeleton inside the hollow bronze form, and a huge external brace hidden by the cape of the Colossus. For further stability each section of the hollow legs was filled with stone as soon as it was in place. (Postwar Rhodes could now reconvert its stone fortifications to peace-time public works.)

Chares built a model of each piece before the bronze part was made. Then workmen would copy his miniature. The

Ptolemy III Eurgetes of Egypt offered to cover all restoration costs for the toppled monument. However, an oracle was consulted and forbade the re-erection. Ptolemy's offer was declined.

most highly skilled craftsmen on the island supervised this work, then personally did the final metalworking for each part.

What went wrong?

At least two historians who later visited the site—Pliny and Phylon—regarded the Colossus as the most perfect execution of the human form they had ever seen. Its architect was less satisfied. Chares committed suicide before the Colossus was finished, and his work had to be completed by Laches.

Above: The harbor of Rhodes now is guarded by two stags.

Opposite: Rhodes is now one of Greece's most beautiful cities—and a cradle of history from the ancient Greeks to the Crusades. The cobbled Street of the Knights runs through one of the oldest parts of the city.

When the Colossus fell, few people could make their arms meet round its thumb.

Accounts differ as to why Chares killed himself, but apparently his reason was one of two common frustrations of architects in all ages. One version has it that Chares had become disconsolate over some flaw in the design. The other is that he thought he had lost face when it became apparent that the costs of the project would be substantially more colossal than his original estimates.

The second version has certain earmarks of a history written by an architect, but it should be remembered that the Rhodians had good reason to expect that the estimate would be accurate. The Colossus of Rhodes was only one—albeit the largest and best one—of roughly a hundred "colossal" statues that populated the island.

The Colossus of Rhodes by Antonio Munoz Degrain takes a colorful look at a dramatic harbor scene—from the perspective of a 20th-century Spaniard.

For almost a millennium, the statue laid broken in ruins. In 654 AD, the Arabs invaded Rhodes. They disassembled the remains of the broken Colossus and sold them to a Jew from Syria. It is said that the fragments had to be transported to Syria on the backs of 900 camels.

There have been many accounts of the final form and function of the Colossus. Some are clearly apocryphal, such as the legend that is stood astride the harbor, with ships coming and going between its legs. Some controversies have yet to be settled—for example, the argument as to whether the Colossus held a harbor light aloft.

The fallen idol

The reason for this uncertainty, and for the scarcity of good descriptions of the Colossus, is that it stood for only fifty-six years before it was felled by an earthquake. It was in this prone and fragmented condition that most chroniclers saw it before writing their accounts.

Pliny reported: "Few men can clasp the thumb in their arms, and its fingers are larger than most statues."

Hardly a century went by without at least one attempt to raise the fallen idol, but all efforts failed. Finally, in A.D. 653, the Saracens carted off the remains—by merchant fleet to the mainland, then inland on the backs of 950 camels—for reconversion into weapons.

The Colossus of Rhodes has a great deal in common with the Statue of Liberty. Many accounts tell of a torch held aloft, although historians have not been able to explain a method to keep it burning.

Although it disappeared from existence, the ancient world wonder inspired modern artists such as French sculptor Auguste Bartholdi, best known for his famous work, The Statue of Liberty.

"To you, O Sun, the people of Dorian Rhodes set up this bronze statue reaching to Olympus when they had pacified the waves of war and crowned their city with the spoils taken from the enemy. Not only over the seas but also on land did thy kindle the lovely torch of freedom."

—The dedicatory inscription on the Colossus of Rhodes

Index

United States Military Academy, 54, *54*

Ur (ancient city), 182, *187*

Uruk (ancient city), 182

Valhalla, 123, 131

van Cleve, Hendrick, 44-45

Van Verghaeght, Tobias, 38

Vast, Jean, 98

Vaux-le-Vicomte, France, 156-165, *156-161*

vermin, 102

Versailles, France, 158, 159, 162, 164, *164*

Vikings, 29, 32, 130, 131, artifacts, of 123; construction methods of, 123, 125, *125*; graveyards, 126, *126*; ships and shipbuilding, *120*, 123, 125, *126*, 127, 130, 131

Viollet-le-Duc, Eugene, 93

walls, 62, 100-109, 177, 178, 179, 182, 183, 185, 187, 191; construction of, 102, 104, 105, 109; *see also* battlements, fortifications

Wan-Li Qang-Qeng, *see* Great Wall of China

watchtowers, 102, 104, 182, 185; *see also* defense towers

water, 146

weaverbirds, *70, 73, 74, 75*

Whinney, Margaret, 17

wigwams, *132*, 138

Wilcox, William H., 141

Winchester Mystery House, San Jose, 46-53, *46, 47, 49, 50-51, 52, 53*

Winchester, Sarah Pardee, 47-49, *48*

Woden (god), 123, 126, 131

Wotan (god), 131

Wren, Christopher, 12-25, *14;* correspondence of, 20-21

Xerxes, 42

ziggurats, 37, *37,* 40, *187*

Picture Credits

Ancient Art & Architecture Collection Ltd.: p. 12 (Ronald Sheridan), 15 (Ronald Sheridan), 17b (Ronald Sheridan), 26 (Jim Conboy), 28 (Cheryl Hogue), 32 (Jim Conboy), 38 (Ronald Sheridan), 40 (Ronald Sheridan), 43t (Ronald Sheridan), 55b, 58 (J. Stevens), 61 (B. Norman), 63b (Mary Jelliffe), 67b (Mary Jelliffe), 68 (Mary Jelliffe), 81t (Gianni Tortoli), 88t (James Lynch), 101 (William Lindesay), 103 (William Lindesay), 104, 106 (William Lindesay), 109 (William Lindesay), 120 (Leslie Ellison), 133 (Ronald Sheridan), 134 (B. Norman), 146 (Julian Worker), 147b (Mary Jelliffe), 152 (D. Matherly), 154t (B. Norman), 155b (Dr. S. Coyne), 157 (Haruko Sheridan), 159 (Haruko Sheridan), 163, 164 (Ronald Sheridan), 168 (Ronald Sheridan), 169 (Brian Gibbs), 171 (Ronald Sheridan), 187 (B. Norman), 189 (Dr. Alan Beaumont), 192 (Ronald Sheridan), 194 (Dr. Alan Beaumont), 195

Hulton Getty/Archive Photos: p. 13, 29r, 33, 135 (both), 138, 174, 175, 182

Corbis: p. 14 (Chris Andrews), 18 (Bertrand), 27 (Nowitz), 41 (Francis Meyer), 42t (Chris Hellier), 42b, 54, 55t, 56, 62 (Vikander), 66, 69 (Ergenbri), 74b (Tom Bean), 76 (Audubon), 77b (Frank Lane), 96 (Almasy), 111 (Chris Rainier), 112, 118 (Chris Rainier), 130 (Ted Spiegel), 131 (Harper's Weekly), 132 (Angelo Hornak), 160 (A. Woolfitt), 188 (Araldo de Luca), 190t, 191 (Kit Kittle)

Bridgeman: p. 16 (Guildhall Library), 36, 41, 44, 90, 92 (Peter Willi), 93 (Peter Willi), 156 (Peter Willi), 158, 162, 165 (Peter Willi), 167, 173

Edifice/Lewis: p. 17t, 136, 166, 172

Edifice/Darley: p. 145

Edifice/Sayer: p. 22, 30 (Heini Schneebeli), 144

Edifice/C. Mellis: p. 153 (both)

Barnaby's Picture Library: p. 20 (Colin Underhill), 35 (Tuck Goh), 74t (B.E. Enstone), 75 (Gerald Cubitt), 94, 170, 184 (S. Nickels), 186

Still Moving Picture Company: p. 29 (Ken Paterson), 34 (Angus G. Johnston)

AP/World Wide Photos: p. 37, 108 (Greg Baker), 110 (Stephan Savoia), 113 (Stephan Savoia), 197

Sonia Halliday Photographs: p. 39 (Polly Buston), 43b (Laura Lushington), 72 (Bryan Knox Arps), 77t (Bryan Knox Arps), 78 (Jane Taylor), 85 (Jane Taylor), 88b (Jane Taylor), 89b (Jane Taylor), 91, 99(Laura Lushington), 148 (Bryan Knox Arps), 155t (Laura Lushington), 177, 179 (Laura Lushington), 180, 183 (Barry Searle)

Winchester Mystery House: p. 46, 47, 48, 49, 50, 52, 53

National Geographic Society: p. 56 (Jonathan Blair), 59 (Hiram Bingham), 60 (William Allard), 71 (Bates Littlehales), 73l (O. Louis Mazzatenta), 73r (Chris Johns), 83 (Annie Griffiths Belt), 86 (Annie Griffiths Belt), 89t (Annie Griffiths Belt), 102b (James Stanfield), 105t (Sidney Hastings), 105b (James L. Stanfield), 124 (James Stanfield), 125 (Ted Spiegel), 126 (Ted Spiegel), 128 (Ted Spiegel), 143 (Raymond K. Gehman), 154b (Edward S. Curtis), 176 (Richard T. Nowitz)

Royal Geographical Society: p. 63t (John Miles), 67t (Eric Lawrie), 79 (C. Bradley), 80 (Chris Caldicott), 81b (Chris Caldicott), 82 (David Roberts), 84 (Chris Caldicott), 102 (Herbert Ponting), 114 (Worle & Longstaff), 115 (R. Amundsen), 116 (R. Amundsen), 117 (British Arctic Air Route Expedition), 150 (Victoria Keble-Williams)

Photodisc: p. 64, 70, 100

The British Library: p. 122, 123

Werner Forman/Art Resource: p. 147T, 151

Giraudon/Art Resource: p. 121

The Jewish Museum, NY/Art Resource: 178